Introducing Special Relativity

Introducing Special Relativity

W. S. C. Williams

First published 2002
by Taylor & Francis
11 New Fetter Lane, London EC4P 4EE

Simultaneously published in the USA and Canada
by Taylor & Francis Inc,
29 West 35th Street, New York, NY 10001

Taylor & Francis is an imprint of the Taylor & Francis Group

© 2002 Taylor & Francis

Typeset in 10/12 pt Times New Roman by
Newgen Imaging Systems (P) Ltd., Chennai, India
Printed and bound in Great Britain by
TJ International Ltd, Padstow, Cornwall

British Library Cataloguing in Publication Data
A catalogue record for this book is available
from the British Library

Library of Congress Cataloging in Publication Data
A catalog record for this book has been requested

ISBN 0-415-27761-2 (hbk)
ISBN 0-415-27762-0 (pbk)

For
William, Thomas, Samuel and Sophie
Perhaps, one day, one of them might be interested

Contents

Preface   xi
Acknowledgements   xii

1   Introduction   1

2   Momentum and energy in special relativity   4

   **2.1   *Introduction*   *4*
   **2.2   *Momentum and energy*   *4*
   **2.3   *More useful formulae*   *6*
   **2.4   *Units*   *8*
   **2.5   *Rest mass energy*   *9*
   **2.6   *Nuclear binding energy*   *10*
   **2.7   *Nuclear decays and reactions*   *11*
   **2.8   *Gravitation and $E = Mc^2$*   *13*
   **2.9   *Summary*   *14*
             Problems   *15*

3   Relativistic kinematics I   17

   **3.1   *Introduction*   *17*
   **3.2   *Elementary particle physics*   *18*
   **3.3   *Two-body decay of a particle at rest*   *20*
    *3.4   *Three-body decays*   *24*
    *3.5   *Two-stage three-body decay*   *26*
   **3.6   *The decay of a moving particle*   *28*
     3.7   *Kinematics of formation*   *30*
     3.8   *Examples of formation*   *33*
   **3.9   *Photons*   *35*
   **3.10   *Compton scattering*   *35*
   **3.11   *The centre-of-mass*   *37*

**3.12   *The centre-of-mass in two-body collisions   41*

*3.13   *Reaction thresholds in the laboratory   42*

*3.14   *Summary   43*

Problems   43

4   Electrodynamics   46

**4.1   *Introduction   46*

**4.2   *Making contact with electromagnetism   46*

**4.3   *Relativistic mass   49*

**4.4   *Historical interlude   50*

*4.5   *Bucherer   51*

*4.6   *Accelerators   54*

*4.7   *The cyclotron frequency   55*

*4.8   *The cyclotron   56*

*4.9   *The limitations of the cyclotron   57*

4.10   *The synchrocyclotron   58*

4.11   *The synchrotron   59*

4.12   *Electron accelerators   60*

4.13   *Centre-of-mass energy at colliders   60*

*4.14   *Summary   61*

Problems   62

5   The foundations of special relativity   64

**5.1   *Introduction   64*

**5.2   *Some definitions   65*

**5.3   *Galilean transformations   68*

**5.4   *Electromagnetism and Galilean relativity   70*

**5.5   *Einstein's relativity   72*

**5.6   *The rise and fall of simultaneity   72*

**5.7   *Time   74*

**5.8   *Einstein's electrodynamics   76*

**5.9   *Comments on Einstein's results   78*

**5.10   *Charge invariance   80*

**5.11   *Summary   80*

Problems   81

6   Relativistic kinematics II   82

**6.1   *Introduction   82*

**6.2   *The Lorentz transformation of energy and momentum   82*

*6.3   *The Lorentz boost for optics: aberration   87*

*6.4   The relativistic Doppler effect   90

*6.5   Aberration and the Doppler effect in astrophysics   91

6.6   Lorentz boosts and new variables   93

+6.7   Graphical representation of the Lorentz boost   96

*6.8   Lorentz boosts at work in particle decays   99

**6.9   Collision kinematics   100

*6.10   Summary   102

Problems   102

7   The Lorentz transformation of time–space coordinates   109

**7.1   Introduction   109

**7.2   The Lorentz transformation of time–space coordinates   109

**7.3   Conformity with the postulates   111

**7.4   A simplification in notation   112

*7.5   A world line   113

*7.6   The interval   115

**7.7   The absence of simultaneity in special relativity   117

**7.8   The Lorentz–FitzGerald contraction   119

7.9   A diversion   122

**7.10   Time dilation   122

**7.11   The experimental tests of time dilation   125

**7.12   The travelling twin   128

*7.13   Proper time   131

*7.14   Time dilation and the Lorentz–FitzGerald contraction   132

7.15   The transformation of velocities   132

7.16   The composition of velocities   134

+7.17   Minkowski map   136

+7.18   The Lorentz–FitzGerald contraction and time dilation
on the Minkowski map   139

*7.19   Summary   140

Problems   140

8   The formalities of special relativity   145

**8.1   Introduction   145

**8.2   The derivation of the Lorentz transformation   145

8.3   Derivation using the k-calculus   147

**8.4   Derivation using a conventional method   152

**8.5   Derivation of the transformation of the
transverse coordinates   155

*8.6   Summarising the Lorentz transformations   157

8.7   The general Lorentz boost   158

$^+8.8$ *Two successive Lorentz boosts I   159*

$^+8.9$ *Two successive Lorentz boosts II   160*

$^*8.10$ *Four-vectors   164*

$^*8.11$ *A classification of four-vectors   169*

$^+8.12$ *Relativistic energy and momentum conservation validated   169*

$^+8.13$ *The Thomas precession   175*

           Problems   178

9   Some developments in special relativity   182

$^*9.1$ *Introduction   182*

$^*9.2$ *Relativistic kinematics with four-vectors   182*

$^*9.3$ *More on waves   188*

$^*9.4$ *Einstein to de Broglie   190*

$^+9.5$ *Schrödinger's wave equation   192*

$^+9.6$ *The Dirac equation   195*

9.7   *Maxwell's equations   198*

9.8   *The Lorentz transformation of E and B   200*

$^+9.9$ *The covariance of Maxwell's equations   204*

$^+9.10$ *The electromagnetic four-vector potential   206*

$^+9.11$ *Tensorland   208*

$^+9.12$ *Electromagnetism with tensors   215*

$^*9.13$ *What to remember   219*

$^*9.14$ *What we have not discussed   219*

           Problems   220

10   A postponed prelude   226

$^{**}10.1$ *Introduction   226*

$^{**}10.2$ *Maxwell and the ether   226*

$^{**}10.3$ *The hunt for the ether: Michelson and Morley   227*

$^{**}10.4$ *The hunt for the ether: Trouton and Noble,*
               Rayleigh, and Brace   230

$^{**}10.5$ *Saving the ether with Lorentz!   230*

$^{**}10.6$ *Dragging the ether   232*

$^{**}10.7$ *1905 and after   234*

Problem answers   237

Bibliography   239

Index   243

Preface

This book is intended for first or second year university undergraduates studying physics. The objective is to provide an easy and rewarding route into special relativity. The subjects that cause students difficulty, such as time-dilation and the Lorentz–FitzGerald contraction, are delayed to allow the establishment of confidence that the greater part of relativity is easy to understand and to apply correctly. The use of four-vectors is important to this objective. Therefore, the reader is guided gradually towards their use.

The relativistic energy and momentum of a particle and the concept of rest mass energy are introduced early. The meaning of $E = Mc^2$ is discussed and we show how very easy it is to solve problems in relativistic kinematics. The sections following show how Maxwell's very successful theory of electromagnetism is inconsistent with the classical concept of relativity implicit in Newton's mechanics. Einstein's principle of relativity is introduced with a summary of the methods and results of his two 1905 relativity papers. Chapters follow on the application of the Lorentz transformations to energy–momentum and to time–space and the consequences, including, of course, time dilation and the Lorentz–FitzGerald contraction. The remaining chapters cover the derivation of the Lorentz transformations, developments in twentieth-century physics that followed, and a return to electromagnetism to demonstrate that it is uniquely constrained by relativity. The final chapter reviews some of the early experiments relevant to the development of the subject.

I have benefited greatly from colleagues who have been prepared to discuss relativity with me. Among these are Harvey Brown who read a draft and discussed the logic of the derivation of the Lorentz transformation, and Brian Buck who alerted me to the debate over Einstein's derivation of $E = Mc^2$ given in his second 1905 relativity paper. But I am particularly grateful to Ian Aitchison for reading a draft and for some very timely and pertinent advice. If I have failed to understand what they said or if I have failed to follow properly that good advice, or if I am otherwise in error, then the fault is mine.

W. S. C. Williams
Oxford
April 2001

Acknowledgements

I am very pleased to thank the following persons and organisations:

1   Sidney Harris ⓒ for the cartoon of the salesman, a customer and a sports car with steering wheel on the right for use on roads in Britain (illegally at the top speed proposed).
2   The Institute of Physics Publishing Ltd for permission to reproduce (Figure 3.1) Plate I from the article by C. F. Powell which appeared in Reports in Progress Physics, Volume 13, pp. 350–424 (1950).
3   The Lawrence Berkeley National Laboratory for permission and a copy of the bubble chamber photograph used in Figure 3.4.
4   The European Organisation for Nuclear Research (CERN) for permission and a copy of the bubble chamber photograph used in Figure 6.5.
5   The American Astronomical Society, W. N. Colley and E. L. Turner (Princeton University), J. A. Tyson (Bell Labs, Lucent Technology), NASA/STScI, for permission to use the cover photograph.

I am also, once again, very grateful to Irmgard Smith who prepared all the line drawings. Thanks also to the staff at Taylor & Francis (and Gordon and Breach as was) for guidance and help.

Cover photograph information

This photograph is a tribute to Einstein. At the centre is a cluster of elliptical and spiral galaxies (Cl 0024 + 1654 in the constellation of Pisces). In these galaxies, about a million million stars are using nuclear reactions to convert mass into energy. The exchange rate is given by Einstein's $E = Mc^2$, a result from the theory of special relativity. Look for the ring of blue knots centred on the cluster. They form a fragmented Einstein ring. Each knot is an image of the same, single, more distant galaxy which is behind the cluster. It is 'lensed' by the gravitational field of the cluster. This lensing gives brightened and distorted images of that galaxy. The blue colour indicates a galaxy that is emitting light in the ultra-violet (wavelength 160 to 205 nm) which is 'red-shifted' to the blue. This, with other evidence, indicates a young galaxy with very active star formation. The gravitational bending of the light is an effect predicted by Einstein's general relativity. The observed red-shift in the space–time of the expanding Universe is the generalisation of the Doppler shift of light from a receding source in special relativity. The cluster has a red-shift, $z = 0.39$, and is about 5 thousand million light years distant. The lensed galaxy is more than twice as far away.

The photograph was taken by the Hubble Space Telescope. For details see:

Colley, W. N., Tyson, J. A., and Turner, E. L. (1996) *Astrophysical Journal*, **461**, L83–L86.

"ITS TOP SPEED IS 186 M.P.H. — THAT'S ⅓,600,000 THE SPEED OF LIGHT."

1 Introduction

In his book *A Brief History of Time*, Stephen Hawking tells how he was told that every algebraic equation included in his text would halve his sales. He took a risk, included one, and produced a best seller. The equation, irrevocably linked to the name of Albert Einstein, is $E = Mc^2$. Its rise to public prominence in the twentieth century came with the development of nuclear weapons and of nuclear power. We shall meet it soon, but there will be a few others, although the number to be understood and remembered is only six! We enumerate those in Section 9.13.

At the time of writing the twentieth century is drawing to a close. A hundred years ago physicists were tackling the unsolved problems of classical physics. A major one concerned the reconciliation of the foundations of classical mechanics with the very successful theory of electromagnetism developed by James Maxwell. Among many distinguished physicists involved in this endeavour, the names George FitzGerald, Joseph Larmor, Henri Poincaré, Hendrik Lorentz are prominent. But it took the genius of Einstein to bring simplicity where previously there had been strained hypotheses. He modified classical mechanics in a manner which was to be essential to the development of modern physics. Thus was **Special Relativity** born.

The effects the theory describes are particularly important for bodies moving with speeds that are a significant fraction of the speed of light. Such speeds are sometimes labelled as relativistic. Before special relativity, the theory of light was based on the presumed existence of the *ether*, the medium classical physicists needed to support electromagnetic waves. Experiments designed to detect the existence of this ether had failed. Einstein's relativity explained this failure and solved the problems associated with electromagnetism, without requiring an ether of any kind. Thereafter that concept withered away. Theoretical physicists had embraced special relativity as correct by the time Pauli published an encyclopaedia article on both special and general relativity in 1921.[1] However, apart from the experiments that failed to detect the ether, precise verification of the predictions of special relativity came slowly, although surely. The investigations of the systematics of

---

1 English translation: Pauli, W., *Theory of relativity*, New York, Dover Publications: 1981.

nuclear masses verified $E = Mc^2$ during the decade preceding 1940. The systematic study of particle collisions at relativistic velocities was made possible by the development of high-energy particle accelerators. These collisions could only be analysed successfully and correctly by the use of the kinematics of special relativity. Relativistic quantum mechanics is now an essential tool in the quantitative description of many phenomena in physics, particularly in atomic and sub-atomic physics.

Special relativity made some predictions that many found hard to accept. These included the FitzGerald–Lorentz contraction of the length of a moving body, the slowing down of a moving clock, and the so-called twin paradox. These topics were still the subject of correspondence in the columns of a distinguished scientific journal fifty years after Einstein published his first two papers on relativity in 1905. Therefore it is not surprising that many students find it hard to reach an understanding of the subject. The equations that give these hard-to-believe and challenging results are easy to write down but equally easy to use incorrectly. They describe situations so remote from everyday experience that we find it hard not to be misled by many ingrained, but incorrect, ideas. And the jargon used to label the effects can confuse. For example, the FitzGerald–Lorentz *contraction* of the length of a moving body is what a stationary observer will deduce from the only sensible prescription for the measurement of such a length. For an observer moving with the body, to whom it will appear at rest, there is no contraction.

In view of the difficulties that special relativity presents and to encourage a growing confidence in the reader, we discuss the easiest things first. One of the easiest is using the formulae for energy and momentum of a particle moving at a relativistic speed plus the familiar conservation of energy and momentum to analyse elementary particle decay and collision processes. Therefore in Chapter 2 we introduce the formulae for the energy and momentum of a particle, without justification. One of the results that we recognise is that a particle of mass M at rest has energy Mc^2. The consequences of this are examined in four sections of this Chapter 2. In Chapter 3 we give a series of examples concerning relativistic conservation of energy and momentum, and introduce the idea of the centre-of-mass and of reaction thresholds. Chapter 4 continues the policy of building on the easy material. We make contact with electromagnetism. Since we have used the formula for the momentum of a particle, we can introduce the formula for the rate of change of momentum in electric and magnetic fields. This allows us to examine the early tests of Einstein's relativity and to follow the effect of relativity on the development of accelerators and colliders. That makes Chapter 5 the point at which we have to start to study the real business of relativity. We examine the progression from Galileo's relativity to Einstein's relativity and map out how from two postulates he laid the foundations of special relativity. And we show how his results give the formulae for the relativistic energy and momentum that we use in Chapters 3 and 4. All that leaves the reader ready, we trust, to tackle the Lorentz transformations.

A Lorentz transformation connects the values of the time plus space coordinates of an event in two different coordinate systems which are in relative motion. It was for this role that Einstein showed how to derive the equation of the transformation from the principles of relativity. But that is a particular task. The same equation will act to connect, between the same two systems, the values of a whole class of quantities other than the time plus space coordinates. The members of this class are called four-vectors. The reader will be properly introduced to these objects later. At this moment we note that application to the time plus space coordinates leads to the counter intuitive results that many found hard to accept. For this reason, in Chapter 6, we examine the transformation of the energy and momentum of a particle, and its consequences. The results are physically reasonable, have been verified and should not challenge anyone's intuition. That challenge we leave to Chapter 7. There we explore the properties of the transformation of the time–space coordinates.

In Chapter 8 we present a derivation of the Lorentz transformation of the time–space coordinates, take a look at some of its properties and give a formal justification for the formulae used in earlier chapters. In Chapter 9 we discuss some of the developments in physics that were made possible by relativity. We also go back to electromagnetism. Of course, it was problems with electromagnetism that prompted Einstein. However, many undergraduate courses now introduce relativity before the students have reached a good understanding of Maxwell's equations. Therefore the route taken by Einstein in his first 1905 relativity paper can be fully appreciated only by the more advanced students. We use Maxwell's equations to show that electromagentism is a relativistic subject and that this theory may be reformulated in a way that is manifestly and properly relativistic. In the final chapter (10) we discuss some of the experimental observations of the nineteenth century which indicated the failure of some classical ideas and which were a prelude to the birth of special relativity.

In the table of contents and at the section headings we have marked, with ** or *, those sections that could form the basis of an introduction to relativity for student beginners in university physics. The ** indicates a sub-group of * for a shortened version of such a course. Conversely, those sections that would be suitable for the more advanced students have been marked +. Unmarked sections are in-between material or concern subjects of less than primary importance. Chapter 10 covers material often found at the beginning of books on relativity and may be read at any time.

2 Momentum and energy in special relativity

**2.1 Introduction

In this chapter we introduce the relativistic expressions for the momentum and energy of a particle. The reader should note that the adjective *relativistic* that we have just used has come to mean a qualification that pertains to Einstein's special relativity or general relativity although before Einstein, a principle of relativity was introduced by Galileo (Section 5.1). Thus a relativistic speed is one that is a significant fraction of the speed of light. And what do we mean by *particle*? Strictly it is an object with mass but without size or shape, sometimes called a *point mass* (Section 5.2). In Newtonian mechanics many bodies that are extended may be approximated by particles. In modern physics the word *particle* is applied to the atomic and sub-atomic constituents of matter (Section 3.2).

We delay justifying these momentum and energy expressions until Section 8.10. Proving that they are the quantities that contribute to the conservation of momentum and energy in particle collisions and related processes will be considered in Section 8.12.

**2.2 Momentum and energy

We need to define **kinematics**. It is the study of the motion of particles or of bodies without reference to the forces acting between them. These forces may lead to collisions between particles in which energy and momentum might be transferred. None the less total momentum and total energy are conserved overall from the time when the particles are free of the forces before the collision to the time after the collision when the particles are again free. Thus a kinematic analysis relates the before and after momenta and energies.

One application of special relativity is to the kinematics of collisions between particles moving with speeds that are a substantial fraction of the speed of light and to processes in nuclear and elementary particle physics. One consequence of Einstein's results is that mass has energy. The equation is the famous $E = Mc^2$,

---

Every heading with ** or * indicates a section we suggest is included in a first course in special relativity. The ** indicates sections for a shortened version of such a course.

where M is mass and c is the speed of light. This means that mass can be transformed into kinetic energy and electromagnetic energy, and, in collisions, kinetic energy or electromagnetic energy can become the mass of created particles.

Most readers will be familiar with the use of conservation of energy and momentum to work out the relation between angles and energies that are the consequence of elastic collisions between billiard balls, for example. Extending those principles to collisions and other processes involving atomic, nuclear and sub-nuclear particles at relativistic speeds is a straightforward matter. We take advantage of a study of this subject to provide the reader with a painless introduction to special relativity.

We start by listing some of the familiar properties of pre-Einstein particle mechanics as would apply in elastic collisions between particles:

1 Momentum is conserved.
2 Energy is conserved.
3 The momentum, P, of a particle is given by

$$P = Mv,\qquad(2.1)$$

where M is again the mass of the particle and v is its speed.
4 The kinetic energy T of a particle is given by

$$T = \tfrac{1}{2}Mv^2.\qquad(2.2)$$

5 Mass is conserved.

Try Problem 2.1 to check that we agree about the conservation of energy and momentum in pre-Einstein mechanics.

In special relativity these properties are partly modified. Properties 1 and 2 are retained. Properties 3 and 4 appear in a new form and 5 has to be replaced by a completely new statement concerning the role of mass:

3′ The momentum of a particle is given by

$$P = \frac{Mv}{\sqrt{1 - v^2/c^2}},\qquad(2.3)$$

where c is the speed of light.
4′ The kinetic energy of a particle is given by

$$T = Mc^2 \left[\frac{1}{\sqrt{1 - v^2/c^2}} - 1 \right].\qquad(2.4)$$

5′ In his second 1905 paper on special relativity Einstein showed that if a body gives off energy L in the form of radiation its mass diminishes by L/c^2. (However, see Section 5.8 for a remark about this result.) This is the basis of the conclusion that mass and energy are equivalent and that the energy content of a body of mass M is Mc^2. This energy can be interchanged with kinetic or electromagnetic energy and therefore mass is not necessarily conserved in collisions or in related processes. Any changes of mass are balanced by corresponding changes in kinetic energy or in electromagnetic energy. Conservation of energy now extends to include this mass energy of all the particles that are involved. We are now using M to mean the mass of a particle at rest, so that its **rest mass energy** is Mc^2. Each particle has a total energy E equal to its rest mass energy plus its kinetic energy:

$$E = Mc^2 + T = \frac{Mc^2}{\sqrt{1 - v^2/c^2}}. \tag{2.5}$$

We wish to reserve the phrase *total energy* for another quantity so by the **energy** of a particle we shall mean E. If we need to use the kinetic energy it will be named as such and be represented by a T. It is now the energy E of a particle, rather than T, that is counted in the overall conservation of energy in collisions and other changes.

For $v \ll c$ we can expand the reciprocal square root in equations (2.3) and (2.5) using the binomial expansion:

$$P = Mv \left\{ 1 + \tfrac{1}{2}(v^2/c^2) + \cdots \right\},$$
$$T = \tfrac{1}{2}Mv^2 \left\{ 1 + \tfrac{3}{4}(v^2/c^2) + \cdots \right\}$$

which tend to the non-relativistic equations, equations (2.1) and (2.2), as $(v/c)^2 \to 0$.

The equation $E = Mc^2$ is widely quoted and used. It is important to remember that this is the equation for the rest mass energy. Given the meaning of our notation it is the energy when $v = 0$. We shall consider some of its simple implications in Sections 2.4–2.7.

The fact that P and $T \to \infty$ as $v/c \to 1$ is taken to mean that no particle with non-zero rest mass may be accelerated to the speed of light. In addition, this result is now taken to mean that special relativity does not permit information to be conveyed between two space points faster than the speed of light (Section 7.6).

**2.3   More useful formulae

We can make some additions to our energy and momentum equations.

1   Eliminate the speed v from the energy (equation (2.5)) and momentum (equation (2.3)) expressions. We find

$$E^2 - P^2c^2 = M^2c^4. \tag{2.6}$$

This is a very important equation. Remember M is the mass of the particle at rest and therefore always the same. We shall use this equation very frequently.

2   Eliminate that awkward factor $\sqrt{1 - v^2/c^2}$ from equations (2.5) and (2.3). We find

$$v/c = Pc/E. \tag{2.7}$$

This is another very important and useful equation.

3   What about photons? They have $v = c$ so equation (2.7) suggests that for these particles

$$E = Pc. \tag{2.8}$$

Now Planck and Einstein taught us that the energy of a photon is given by

$$E = hf,$$

where f is the classical frequency and h is Planck's constant. To this de Broglie added the relation between momentum and wavelength (Section 9.4):

$$P = h/\lambda.$$

Using equation (2.8) to eliminate E and P gives

$$\lambda f = c.$$

This is the usual relation between wavelength, frequency and phase velocity. The phase velocity for light is equal to its group velocity in a vacuum. Therefore equation (2.8) gives us a relation between energy and momentum that is consistent with the physics of photons as the quanta of electromagnetic radiation. For applications and more details of these relations, see Sections 3.9 and 3.10 and Sections 9.3 and 9.4.

4   It is convenient to define a new symbol for the **fractional (of c) speed**:

$$\beta = v/c, \tag{2.9}$$

and to introduce another that will be used frequently:

$$\gamma = \frac{1}{\sqrt{1 - v^2/c^2}} = \frac{1}{\sqrt{1 - \beta^2}} \quad \text{(the Lorentz factor).} \tag{2.10}$$

Then it follows that for a particle of speed v

$$P = \gamma \beta M c \quad \text{and} \quad E = \gamma M c^2. \tag{2.11}$$

5   The $\beta = v/c$ is so useful that we may sometimes carelessly refer to β as the speed rather than as the *fractional (of c) speed*. When these symbols are used it will normally be clear to which speed they apply. But sometimes we

have situations involving more than one speed in which case we shall use, for example

$$\beta_v = v/c, \qquad \beta_u = u/c,$$

$$\gamma_v = \frac{1}{\sqrt{1 - \beta_v^2}}, \qquad \gamma_u = \frac{1}{\sqrt{1 - \beta_u^2}}, \qquad \text{and so on.}$$

However, we shall frequently remind readers about these definitions. Now try Problems 2.2–2.4.

**2.4   Units

We need to remind or to inform readers of the units that are used in atomic and sub-atomic physics. The kinetic energy gained by a particle, of charge q, moving from a point at which the electrostatic potential is V volts to a point at which the electrostatic potential is zero is qV. (The charge q is in coulombs (C) and for convenience we consider a positive charge. Energy is in joules (J).) If the particle is a proton then q is equal to the magnitude of the charge on the electron, e, that is

$$q = |e| = 1.6022 \times 10^{-19}\,\text{C}.$$

If $V = 1\,\text{V}$ we have

$$qV = 1.6022 \times 10^{-19}\,\text{J}.$$

This is an energy of 1 electron volt, symbol eV. In nuclear physics the most appropriate and useful unit is the megaelectron volt (1 MeV $= 10^6$ eV). In particle physics it is the megaelectron volt or the gigaelectron volt (1 GeV $= 1000$ MeV).

The relation (equation (2.6)) between (total) energy E, momentum P and mass M is

$$E^2 - (Pc)^2 = (Mc^2)^2.$$

This means that E, Pc and Mc^2 all have the dimensions of energy. P is therefore often given in units of GeV/c, or MeV/c. Similarly M is frequently given in units of MeV/c^2 or GeV/c^2. Many sources of data call the rest mass energy the mass, and give, for example, the mass of the proton $M_p = 938.27$ MeV. What is meant is that $M_p = 938.27$ MeV/c^2. Since we shall be keeping all factors of c in place, all the formulae we shall derive will contain mass terms that are in the form Mc^2, or may easily be put into that form. If a numerical calculation has to be done, such a term is replaced with the rest mass energy, 938.27 MeV in the case of the proton.

For the reader who may wish to make contact with more familiar units involving kilograms, metres, and seconds here is the procedure. Consider the rest mass energy of the proton:

$$M_p c^2 = 938.27\,\text{MeV}.$$

Then

$$M_p = 938.27 \, \text{MeV}/c^2$$
$$= 938.27 \times 10^6 \times 1.60218 \times 10^{-19} \, \text{J}/c^2$$
$$= 1.50328 \times 10^{-10}/(2.99792 \times 10^8)^2 \, \text{kg}$$
$$= 1.67262 \times 10^{-27} \, \text{kg}.$$

Similarly for momentum we might have

$$Pc = 700 \, \text{MeV}.$$

Then

$$P = 700 \times 10^6 \times 1.60218 \times 10^{-19} \, \text{J}/c$$
$$= 1.12152 \times 10^{-10}/2.99792 \times 10^8 \, \text{kg m s}^{-1}$$
$$= 3.74100 \times 10^{-19} \, \text{kg m s}^{-1}.$$

These two numerical values of M_p and P in SI units (Système International d'Unités) are correct but they do not instantly show that $M_p c^2$ and Pc are comparable in energy units.

This is summarised in Table 2.1. We have included the values of the charge on the electron and the speed of light as used in the calculations we have just completed.

**2.5   Rest mass energy

The discovery of several systems with rest mass that convert their entire rest mass energy into radiation is important support for special relativity. Here are two examples.

1. The **positron** is a positively charged version of the electron (Section 9.6); both have the same mass, $M_e = 511 \, \text{keV}/c^2$. (We use a subscript on the capital M to identify the particle concerned.) Energetic positrons are easily produced and if brought to rest in any material, each will meet an electron forming a hydrogen atom-like bound system. It is electrically neutral and is called **positronium**. It has a binding energy of 6.8 eV. Thus it has a rest mass energy 6.8 eV short of $2 \, M_e c^2 \, (= 1022 \, \text{keV})$. Mutual **annihilation** follows:

$$e^+ + e^- \rightarrow \gamma + \gamma.$$

The photons (γ) have an energy measured to be 511 keV each, with equal and opposite momentum. This is as expected from energy and momentum conservation in a reaction in which a massive object (positronium atom) converts all its rest mass energy into two quanta of electromagnetic energy.

Table 2.1 Units in atomic and sub-atomic physics

Charge on the electron $e = -1.60218 \times 10^{-19}$ C
Speed of light $c = 2.99792 \times 10^8$ m s^{-1}
Planck's constant $h = 6.62607 \times 10^{-34}$ J s
$\hbar = h/2\pi$

Quantity	SI units	Atomic and sub-atomic units (*Based on the electron volt,* eV)
Energy	Joule (J)	$1\,\text{eV} = 1.6022 \times 10^{-19}$ J $1\,\text{keV} = 10^3$ eV $1\,\text{MeV} = 10^6$ eV $1\,\text{GeV} = 10^9$ eV $1\,\text{TeV} = 10^{12}$ eV
Mass	Kilogram (kg)	$1\,\text{eV}/c^2 = 1.7827 \times 10^{-36}$ kg
electron M_e proton M_p neutron M_n pion (π^+ or π^-)M_π pion (π^0)	9.10938×10^{-31} kg 1.67262×10^{-27} kg 1.67493×10^{-27} kg	$0.511\,\text{MeV}/c^2$ $938.27\,\text{MeV}/c^2$ $939.57\,\text{MeV}/c^2$ $139.57\,\text{MeV}/c^2$ $134.98\,\text{MeV}/c^2$ $\uparrow$ These numbers give the rest mass energy immediately. For example: $M_p = 938.27\,\text{MeV}/c^2$ $M_p c^2 = 938.27\,\text{MeV}$
Momentum	kg m s^{-1}	$1\,\text{eV}/c = 5.3443 \times 10^{-28}$ kg m s^{-1} A momentum $P = 100\,\text{MeV}/c$ has $Pc = 100\,\text{MeV}$.

2. The neutral pion is readily produced in, for example, sufficiently energetic proton–proton collisions. It also decays into two photons:

$$\pi^0 \; \rightarrow \; \gamma + \gamma.$$

Now the rest mass of this pion is $135\,\text{MeV}/c^2$. Conservation of energy and momentum requires each photon to have an energy of 67.5 MeV if the π^0 were at rest. However these particles cannot be observed at rest and measurements have to be made on photons from neutral pions moving in the laboratory. But these measurements can be transformed to give the energies as they would be if the π^0 were at rest. The result confirms our expectation.

**2.6 Nuclear binding energy

By 1930 the existence of atomic isotopes was known and accurate measurements had been made of the atomic masses of many isotopes throughout the periodic table, using the techniques of mass spectroscopy. The precision reached allowed the conclusion that each atomic mass was less than the total mass of its presumed constituents. In 1932 Chadwick discovered the neutron and the nature of the constituents clarified: Z protons and $(A - Z)$ neutrons in a nucleus with Z extra-nuclear electrons. Z is the integer atomic number and A the integer atomic mass number. The binding energy $B(Z, A)$ is the energy required to dismantle the atom into its constituents. For stability it must be positive. This energy has mass $M = B(Z, A)/c^2$. Then the separated constituents must have more mass than the atom, as observed. The nuclear share of B is much, much greater than the part due to the binding of the electrons. For helium the numbers are 28 MeV as against 79 eV. Thus to a good approximation, the atomic mass of the atom (Z, A) is (including Z electrons) given by

$$M(Z, A) = ZM_H + (A - Z)M_n - B(Z, A)/c^2,$$

where M_H and M_n are the masses of the hydrogen atom and the neutron respectively.

Nuclear binding energy is difficult to measure directly except for deuterium ($Z = 1$, $A = 2$). This nucleus (called the **deuteron**, symbol d) is a bound state of one proton (p) and one neutron (n). Its binding energy may be determined from energy and momentum conservation in the disintegration reaction caused by a high energy photon (γ) of known energy:

$$\gamma + d \ \rightarrow \ p + n.$$

The result is 2.15 MeV. This is about 0.1% of the rest mass energy of the deuteron. Atomic mass spectroscopy can measure the masses of the proton and the deuteron but not the mass of the neutron. This method provided the first accurate measurement of the neutron mass, assuming $E = Mc^2$, but did not verify this equation.

**2.7 Nuclear decays and reactions

The process of alpha(α)-decay is that in which an atomic nucleus spontaneously emits a nucleus of a helium atom (frequently called an α-particle). This process is found among the naturally occurring radioactive materials. There is a line spectrum of kinetic energies of the emitted α-particles that is unique to each active nuclear species. These energies are between 4 and 9 MeV. An example is:

$$^{238}_{92}U \rightarrow {}^{234}_{90}Th + {}^{4}_{2}He + Q,$$

where $Q = 4.27$ MeV. (For complex nuclei the notation is ^A_ZChemical symbol.) Q is the kinetic energy available from this particular decay to be shared between the emitted α-particle and the recoiling *daughter* nucleus which is the other product of the decay. Where does this energy come from? Of course, it comes from the decrease in mass from the *parent* nucleus ($^{238}_{92}U$ in this case) to the *daughter* ($^{234}_{90}Th$)

plus the α-particle ($^4_2$He). That is

$$Q = [M(Z, A) - M(Z - 2, A - 4) - M(2, 4)]c^2 \tag{2.12}$$

($Z = 92$, $A = 238$ in this case). In terms of binding energies this may be written

$$Q = B(2, 4) + B(Z - 2, A - 4) - B(Z, A)$$

$B(2, 4)$ is approximately 28 MeV. For α-decay to be possible Q must be greater than zero. This is the case for Z greater than about 62 but α-decay is not a significant property of these nuclei until Z is greater than 82.

The division of the energy Q between the α-particle and the daughter nucleus may be calculated using non-relativistic kinematics since in these decays the speeds involved are a small fraction of the speed of light. However, in the next chapter we shall examine decays in which the value of Q is large enough to give the decay products speeds that are a significant fraction of the speed of light. Relativistic kinematics is then essential for a correct analysis of the decay.

All other nuclear decay processes are, of course, also subject to the same rule that energy emitted is accompanied by rest mass loss from initial to final system.

In 1932 Cockcroft and Walton were the first to use accelerator produced energetic particles to cause nuclear reactions. Specifically they studied the reaction

$$p + {}^7_3\text{Li} \quad \rightarrow \quad {}^4_2\text{He} + {}^4_2\text{He}$$

(p stands for proton which is $^1_1$H). They found that there was 14.3 ± 2.7 MeV gain in kinetic energy from the initial to the final system. Within error, this was equal to the decrease in total mass from the initial to the final system, multiplied by c^2.

Let us look at the consequence of building on this success. Consider a general nuclear reaction caused in a collision:

$$a + b \quad \rightarrow \quad 1 + 2 + 3 + \cdots .$$

We define

$$Q = c^2(M_a + M_b - M_1 - M_2 - M_3 - \cdots), \tag{2.13}$$

T_i as the total kinetic energy in the initial state of a and b,

and

T_f as the total kinetic energy in the final state.

Conservation of energy then requires

$$T_f = T_i + Q \tag{2.14}$$

A measurement of T_i and T_f gives Q which in turn provides a relation between the atomic masses and nuclear binding energies involved. Small changes in binding energy may be determined with precision. Thus the edifice of all data from mass spectroscopy, nuclear reactions and nuclear decays allows the atomic masses and nuclear binding energy of individual isotopes throughout the periodic table to be determined with high precision. In 1937 a determination of the speed of light from the then existing data was within 0.5% of the modern value. Sixty years later the data on atomic masses and nuclear energy levels is hugely expanded but it has its secure foundation in Einstein's mass–energy relation.

It is worth noting that, as with α-decay, most nuclear reactions involve low ($<20\,\text{MeV}$) kinetic energies. This means that if nucleons (neutrons and protons) and heavier ions are involved the speeds are a small fraction of the speed of light. Thus the analysis of reactions does not normally require relativistic kinematics. Conservation of momentum may be applied using Newton's momentum equation (2.1). Kinetic energy, equation (2.2), may be used, but in the energy conservation equation, any non-zero Q-value must be included. This is a kind of hybrid kinematics justified if all the velocities of non-zero rest mass particles are small compared with c. Photons can participate in this simplification if each is taken to have a momentum $P = E/c$. Try Problems 2.6 and 2.7. The reader will correctly anticipate that we shall examine the kinematics of decays where Q is sufficiently large that the speeds of the decay products have to be treated relativistically (Chapter 3).

**2.8   Gravitation and $E = Mc^2$

The mass that is used in Newton's Second Law, the mass that appears in the relativistic energy–momentum relations, the mass that has figured in our discussions to this point, is the **inertial mass**. The mass that appears in Newton's law of gravitational attraction is the **gravitational mass**.

Newton introduced what is now known as the *weak equivalence principle* (WEP). For our immediate purposes it means that all bodies of equal mass experience the same acceleration due to a gravitational field independently of their internal structure or composition. This has been checked to a precision of 1 part in 10^{11} for aluminium and gold and 1 part in 10^{12} for aluminium and platinum. We can take it to be true for all materials.

Experimentally an atom of ^{195}Pt has less inertial mass than that of its separated constituents, 78 protons, 78 electrons and 117 neutrons, by an amount equal to the total binding energy of that atom divided by c^2. A similar statement can be made about aluminium. The truth of the weak equivalence principle then implies that the binding energy must be correctly counted in the gravitational mass as it is in the inertial mass. The binding energy reduces both the inertial and the gravitational mass equally from the equal values they would have for the separated constituents. Thus Einstein's $E = Mc^2$ has a meaning in gravitational physics in addition to that which it has in special relativity. Energy has gravitational mass.

To give further substance to this conclusion here are two simple deductions.

1   A box, containing one atom of radon moving slowly between collisions with the interior walls, is weighed. The radon atom decays into an α-particle,

an atom of polonium and two free electrons. The energy of decay (Q in equations (2.12) or (2.13)) is shared as kinetic energy between these products. Provided none of this energy leaves the box, its weight after the decay must be the same as before.

2   A body that is heated increases its internal energy and therefore its weight increases.

Present-day techniques are insufficiently sensitive to confirm these deductions. See Problem 2.8.

**2.9   Summary

In Sections 2.5–2.7 we have examined the consequences of $E = Mc^2$. Now we wish the reader to forget this formula because it is unique to a particle at rest. We remind readers that we have assigned the symbol E to represent the energy of a moving particle. This energy is the rest mass energy *plus* the kinetic energy. Equation (2.5) is the one that has to be remembered for this quantity. We shall

Table 2.2 The Newtonian and relativistic expressions for the kinematic quantities of a single particle. The relativistic relations implied in the lines *1, *2, *3, *4 and *5 are those to be remembered. Here we avoid the use of vector notation. We shall start using vectors in the next chapter and these relations will then appear using vector symbols for P and v. That will remind us to take into account the fact that these quantities have direction as well as magnitude. The relations *1 to *5 are repeated in vector form in Section 9.13

Pre-Einstein	Quantity	Einstein	
M (constant)	Mass		
	Rest Mass	M (invariant)	
v	Speed	v	
Mv	Momentum P	$\dfrac{Mv}{\sqrt{1 - v^2/c^2}}$	*1
	Energy E	$\dfrac{Mc^2}{\sqrt{1 - v^2/c^2}}$	*2
$\frac{1}{2}Mv^2$	Kinetic energy T	$E - Mc^2$	
Some dependent relations are:			
	Rest mass energy Mc^2	$\sqrt{E^2 - P^2c^2}$	*3
P/M	Speed v	Pc^2/E	*4
	Fractional speed β	Pc/E	
	Lorentz factor γ	$\dfrac{1}{\sqrt{1 - v^2/c^2}} = \dfrac{E}{Mc^2}$	
For a photon the essential relation to be remembered is			
		$E = Pc$	*5

reserve *total energy* for the situation where we have two or more particles and we want a phrase to signify the sum of their energies.

In the next chapter we shall begin to examine examples of the use of conservation of energy and momentum in processes involving speeds that are a significant fraction of the speed of light. To help the reader we have summarised in Table 2.2 the formulae needed, including, for the purposes of comparison, the pre-Einstein formulae.

Problems

2.1   A proton moving at a speed which is non-relativistic strikes a stationary proton. The incident proton is elastically scattered. Show that after the collision the angle between the direction of motion of the scattered proton and that of the recoiling proton is 90°. (Elastic scattering here means that the kinetic energy is conserved.)

2.2   Protons are accelerated to have kinetic energy of 100 MeV. Calculate their total energy and momentum in units of MeV and MeV/c respectively. Calculate their speed as a fraction of the speed of light.

Repeat these calculations for protons that have a kinetic energy of 10,000 MeV.

2.3   Find the speed of a particle that has a kinetic energy equal to its rest mass energy.

2.4   Electrons are accelerated in the Stanford Linear Accelerator (SLAC) to a total energy of 45 GeV. (a) What fraction of this is kinetic energy? (b) By what fraction do their speeds fall short of the speed of light?

2.5   Electromagnetic radiation from the Sun arrives at the Earth (above the atmosphere) at a rate of 1400 W m^{-2}. Find the rate of loss of mass of the Sun due to the radiation of this energy in all directions. The Earth–Sun distance is 1.50×10^8 km.

2.6   The nuclear reaction studied by Cockcroft and Walton was

$$p + {}^{7}_{3}\text{Li} \ \rightarrow \ {}^{4}_{2}\text{He} + {}^{4}_{2}\text{He}$$

(see Section 2.7). Calculate:

a   The Q-value for this reaction.
b   The binding energy of the atom of ^{7}Li.
c   The kinetic energy available to share between the helium nuclei in the final state if the reaction is caused by protons with a kinetic energy of 500 keV striking lithium at rest in the laboratory.
d   The momentum of each helium nucleus emerging from the reaction if they share this energy equally.

The modern values of the atomic masses are

Hydrogen	1.007825 u
^{7}Li	7.015999 u
^{4}He	4.002603 u
Neutron	1.008665 u,

where 1 u is the unified atomic mass unit and is equal to 1.660565×10^{-27} kg ($= 931.494$ MeV$/c^2$).

2.7   A stationary nucleus of mass M is in an excited state at an energy E_0 above its ground state. It decays to the ground state emitting a photon of energy E. This energy is less than E_0 by the kinetic energy of the recoiling nucleus. Using non-relativistic mechanics for the recoil nucleus's energy and momentum, show that if $E_0 \ll Mc^2$, the energy E is given by

$$E = E_0 \left(1 - \frac{E_0}{2Mc^2} \right).$$

The first excited state of the nucleus europium ($A = 152$) has an excitation energy of 960 keV. It decays to the ground state emitting a photon. By how much is the energy of this photon less than 960 keV? (Assume the mass of the europium nucleus is A unified atomic mass units as given in Problem 2.6.)

A target of europium is exposed to a beam of photons. What energy above 960 keV must a photon have in order that it may be absorbed by a europium nucleus to produce that nucleus in the excited state?

2.8   Assuming the heat capacity of copper is constant at 420 J kg^{-1} K^{-1}, estimate by how much the weight of 1 kg of copper is greater at 1000°C than at 0°C.

3　Relativistic kinematics I

**3.1　Introduction

An important objective of this chapter is to build the reader's confidence in working on relativistic problems. The equipment required for this step is minimal and is reviewed at the end of this section. However this chapter does cover more material than would be necessary for a first reading.

The algebra of relativistic kinematics is about to become encumbered with symbols c, c^2 and c^4, where, as usual, c is the speed of light. Most professionals make the notation simpler by using units in which $c = 1$. This means that Mc^2 is replaced by M, and Pc by P. Although this leads to simpler equations we shall not do that for two reasons:

1　For beginners to the subject a change to $c = 1$ can be confusing. Many will have been trained to ensure that their equations are dimensionally consistent and will be made uncertain if they find the energy E treated as if it has the same units as the mass M.

2　Later we shall be meeting objects called four-vectors. These are similar to ordinary vectors but have four instead of three components. A four-vector may often be defined by an extra component associated with an ordinary three-vector. That extra component may be a familiar quantity but has different dimensions from the components of the three-vector. That difference can be corrected by factors of c. Thus we can make a four-vector from E and the three components of the momentum vector $\mathbf{P}$ (where E and $\mathbf{P}$ have different dimensions) by explicitly using E and $\mathbf{P}c$ (which have the same dimensions) as the components of what we shall later call the energy–momentum four-vector (Section 8.10). In future we shall always include such factors of c as are needed to give all components of any four-vector the same dimensions while keeping all familiar symbols having their usual dimensional status.

Before we look at some simple kinematic examples let us list the 'equipment' required:

1　(Obvious but let us repeat it.) Use conservation of energy and momentum.

2   Do not use the speed of a particle to define its motion. Use only Pc or E, and M. (To emphasise this we are using lower case v for speed but upper case P, E and M.)

3   Have two of the equations in Table 2.2 ready to use. The first is the relation between energy, momentum (a vector) and mass for a particle which we now write in vector notation:

$$E^2 = \mathbf{P} \cdot \mathbf{P} c^2 + M^2 c^4. \tag{3.1}$$

The second is that for the velocity (a vector) of a particle:

$$\mathbf{v} = \mathbf{P} c^2 / E. \tag{3.2}$$

Remember that the word *mass* means the *mass at rest* and the word *energy* of a particle means rest mass energy plus kinetic energy.

4   We shall need coordinate systems if we are to describe the position of events or the momentum of particles. We call a coordinate system a *reference frame* (Section 5.2). Later we have to be prepared to consider reference frames in relative motion. We have mentioned implicitly a particular frame for a given particle. It is one in which the particle is at rest and we call it the particle's *rest frame*. Of course in that frame other particles may be moving.

5   We have just used vector notation for momentum ($\mathbf{P}$). For those readers who are just coming to grips with the subject of vectors we shall issue occasional reminders. We use bold face ($\mathbf{P}$) to represent a vector which has magnitude P. Thus $\mathbf{P}$ represents a (three-)vector that has, in a Cartesian coordinate system, three components P_x, P_y, and P_z. An equation like $\mathbf{P}_1 = \mathbf{P}_2 + \mathbf{P}_3$ is three equations, one for each of the x, y, and z components of the vectors. And

$$\mathbf{P} \cdot \mathbf{P} = P_x^2 + P_y^2 + P_z^2,$$

$$P = \sqrt{P_x^2 + P_y^2 + P_z^2}.$$

Further vector properties are given in Table 3.1.

**3.2   Elementary particle physics

This is the branch of sub-atomic physics from which we will draw many of our examples and problems. It all started with the discovery of the electron in 1897. Our knowledge and understanding of the subject has expanded dramatically since then, particularly in the last half of the twentieth century. It is the environment in which particles interact, are produced and decay at relativistic speeds. Special relativity is essential to understanding the kinematics.

   Readers must not be uneasy about the names and properties of the elementary particles that provide numerical examples for the relativistic kinematics in this chapter. The particle with a mass just over 200 times the mass of the electron,

Table 3.1 Vector notation

Here we add to the vector notation first used at the end of Section 3.1.

- The magnitude of a vector **P** (components P_x, P_y, P_z) is |**P**| which we write as P.
- The notation **P** · **P** means the *scalar product* of **P** with itself:

$$\mathbf{P} \cdot \mathbf{P} = P^2 = P_x^2 + P_y^2 + P_z^2.$$

- For the *scalar product* of two different vectors **P** and **Q** we have

$$\mathbf{P} \cdot \mathbf{Q} = P_x Q_x + P_y Q_y + P_z Q_z = PQ \cos \theta = \mathbf{Q} \cdot \mathbf{P},$$

  where θ is the angle between the directions of the two vectors.
- In Section 3.6 we have:

$$\mathbf{P}_a = \mathbf{P}_1 + \mathbf{P}_2 + \mathbf{P}_3 + \cdots \equiv \sum_i \mathbf{P}_i.$$

  Then we have used the following notation:

$$\mathbf{P}_a \cdot \mathbf{P}_a \equiv \mathbf{P}_a^2 \equiv \left(\sum_i \mathbf{P}_i \right) \cdot \left(\sum_i \mathbf{P}_i \right) \equiv \left(\sum_i \mathbf{P}_i \right)^2 = P_a^2.$$

- For two particles 1 and 2 with $\mathbf{P}_1 + \mathbf{P}_2 = \mathbf{P}_a$ we have

$$\mathbf{P}_a \cdot \mathbf{P}_a = (\mathbf{P}_1 + \mathbf{P}_2) \cdot (\mathbf{P}_1 + \mathbf{P}_2) = P_1^2 + P_2^2 + 2\mathbf{P}_1 \cdot \mathbf{P}_2 = P_1^2 + P_2^2 + 2P_1 P_2 \cos \theta,$$

  where θ is again the angle between the directions of the vectors $\mathbf{P}_1$ and $\mathbf{P}_2$.
- In Section 4.2 we use the *vector (or cross) product* of two vectors; the result is another vector. The notation is $\mathbf{R} = \mathbf{P} \times \mathbf{Q}$. This equation for **R** is shorthand for the three equations:

$$R_x = P_y Q_z - P_z Q_y, \quad R_y = P_z Q_x - P_x Q_z, \quad R_z = P_x Q_y - P_y Q_x.$$

  This appears to be very complicated but try a simple case: suppose the only non-zero components of **P** and **Q** were P_x and Q_y respectively. This would give one non-zero component to **R** which is $R_z = P_x Q_y$. This bears out the important property of the vector product: that the result **R** is a vector perpendicular to both **P** and **Q**. Note that $\mathbf{P} \times \mathbf{Q} = -\mathbf{Q} \times \mathbf{P}$. Then the magnitude $|\mathbf{R}| = |\mathbf{P} \times \mathbf{Q}| = |\mathbf{P}||\mathbf{Q}| \sin \theta$, where θ is the angle between the directions of **P** and **Q**. The spatial relation of **R**, **P** and **Q** in this example bears out the rule for this operation on a pair of vectors (Boas[1]): **R** is in the direction of advance of a right-handed screw turned so as to carry **P** in a sense that reduces the angle between **P** and **Q**.
- Equations (4.2) and (4.3) express physical laws correctly using vector products. Thus from equation (4.2) we deduce that an element of a current circuit placed in the direction the x-axis and in a magnetic induction **B** in the y direction, experiences a force in the z direction of magnitude $I \, dl \, B$
- That is the extent of our need of vectors until Section 9.7 where we quote Maxwell's equations. At that point we have to use the differential vector operators grad, div, and curl. We trust that by then the reader will have become familiar with these operators.

---

1 Boas, M. L., *Mathematical methods in the physical sciences*, 2nd edn, New York: John Wiley & Sons, 1983, pp. 102–3.

which was discovered in the cosmic radiation at sea level in 1936, was named the *mesotron* (meso – implying intermediate in mass between electron and proton). In 1947 the discovery of a second particle of similar mass meant two names, μ-mesotron and π-mesotron! Over the years these names were diminished firstly to mu-meson and pi-meson, then to muon and pion. None the less *meson* lives on (D-meson in Section 3.5) and it now means a particle that experiences the strong nuclear forces and has spin angular momentum that is an integer number of units of $\hbar$ (Planck's constant divided by 2π).

It is worth bearing in mind that in experimental particle physics there are many technical limitations. For example, the momentum of a charged particle may be measured by finding the curvature of its trajectory in a magnetic field (Section 4.2). However, the energy of a relativistic particle is hard to measure with precision. Therefore the identification of a moving charged particle by a determination of its mass from separate measurements of P and E alone is rarely an option. The precise measurement of velocity is sometimes possible, if it is not too close to c. Such difficulties are worse for a neutral particle. Short-lived particles (average lifetime $<10^{-9}$ s) may be identified only by means of a kinematic analysis of decay products, which again depends on measurements of momentum and energy. Examples of these methods will be examined in Sections 3.5 and 3.6.

**3.3   Two-body decay of a particle at rest

We ask readers to go back for a few moments to Section 2.7 on *Nuclear decays and reactions*. We dealt with α-decay, a process in which a nucleus spontaneously emits an α-particle. The kinetic energy in the final state is provided by the decrease in rest mass from the parent nucleus to the daughter nucleus plus α-particle. This energy is small and no particle is moving with a speed as great as $0.1c$. Thus conservation of Newton's momentum can be used to determine how the energy is divided between the α-particle and the recoiling daughter nucleus. We must now deal with decays in which the energy released causes speeds among the decay products that are a significant fraction of the speed of light.

We consider the decay of particle 'a' into two particles 1 and 2 in the rest frame of 'a'. We symbolise this by

$$a \rightarrow 1 + 2.$$

Now let us summarise our notation in a table.

	a	$\rightarrow$	1	+	2
Mass	M_a		M_1		M_2
Energy	$E_a = M_a c^2$		E_1		E_2
Momentum $\times c$	$P_a c = 0$		$P_1 c$		$P_2 c$
			$= \sqrt{E_1 - (M_1 c^2)^2}$		$= \sqrt{E_2 - (M_2 c^2)^2}$

Now

$$Q = (M_a - M_1 - M_2)c^2.$$

The decay is energetically allowed if $Q > 0$ (as in nuclear α-decay in Section 2.7). Assuming that condition is satisfied let us apply the conservation rules:

Conservation of momentum (using vector notation):

$$\mathbf{P}_a = 0 = \mathbf{P}_1 + \mathbf{P}_2. \tag{3.3}$$

Conservation of energy:

$$M_a c^2 = E_1 + E_2. \tag{3.4}$$

From equation (3.3)

$$P_1 = P_2.$$

Therefore, using equation (3.1)

$$E_1^2 - (M_1 c^2)^2 = E_2^2 - (M_2 c^2)^2.$$

Now eliminate E_2 using (3.4):

$$E_1^2 - (M_1 c^2)^2 = (M_a c^2 - E_1)^2 - (M_2 c^2)^2.$$

Solving for E_1 gives

$$E_1 = \frac{(M_a c^2)^2 + (M_1 c^2)^2 - (M_2 c^2)^2}{2 M_a c^2}. \tag{3.5}$$

The kinetic energy of particle 1 is then

$$T_1 = E_1 - M_1 c^2.$$

E_2 and T_2 may be found by exchanging the subscripts 1 and 2. There is a formula for the momentum in the final state:

$$Pc \equiv |\mathbf{P}_1|c = |\mathbf{P}_2|c$$
$$= \frac{c^2}{2 M_a} \sqrt{[M_a^2 - (M_1 + M_2)^2][M_a^2 - (M_1 - M_2)^2]},$$

but this is not interesting since it is easier to calculate P from

$$(Pc)^2 = E_i^2 - (M_i c^2)^2, \quad \text{where } i = 1 \text{ or } 2. \tag{3.6}$$

Use equation (3.2) to find the fractional speeds:

$$\beta_i = Pc/E_i, \quad i = 1, 2. \tag{3.7}$$

Example: The positive pion can decay into a positive muon and a neutrino:

$$\pi^+ \rightarrow \mu^+ + \nu.$$

Given the rest masses are 139.6, 105.7 and 0 MeV/c^2 respectively, calculate the energy, kinetic energy, momentum and speed of each decay product. We use subscript π for the pion, μ for the muon, and ν for the neutrino. Equation (3.5) gives

$$E_\mu = \frac{[(M_\pi c^2)^2 + (M_\mu c^2)^2 - (M_\nu c^2)^2]}{2 M_\pi c^2}$$

$$= [(139.6)^2 + (105.7)^2]/(2 \times 139.6)$$

$$= 109.8 \, \text{MeV}.$$

Equation (3.6) gives

$$Pc = 29.8 \, \text{MeV} \quad \text{or} \quad P = 29.8 \, \text{MeV}/c.$$

Equation (3.7) gives

$$\beta_\mu = Pc/E_\mu = 0.271,$$

so that

$$v_\mu = 0.271 \times 3.00 \times 10^8 = 8.13 \times 10^7 \, \text{m s}^{-1}.$$

Now we may calculate the kinetic energy of the muon:

$$T_\mu = E_\mu - M_\mu c^2 = 4.1 \, \text{MeV}.$$

Since $M_\nu = 0$ (or possibly very small and negligibly so in this problem) the kinetic energy for the neutrino is the same as its energy:

$$E_\nu = Pc = 29.8 \, \text{MeV},$$

and its fractional velocity

$$\beta_\nu = Pc/E_\nu = 1,$$

as it must be for a particle with zero mass (assumed).

In Figure 3.1 we show photomicrographs of positive pion decays at rest recorded in photographic emulsion.

Now try Problem 3.1.

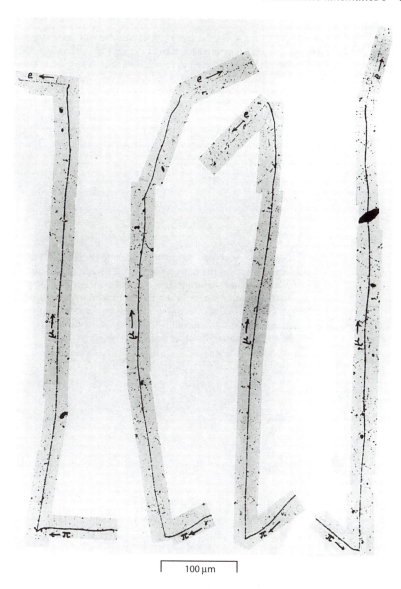

100 μm

Figure 3.1 Photomicrographs of four events in each of which a positive pion (π^+) and a positive muon (μ^+) are observed to decay. These events were found, after development, in the emulsion of special photographic plates exposed to cosmic radiation at high altitudes.[1] Each shows an example of the decay chain

$$\pi^+ \rightarrow \mu^+ \rightarrow e^+.$$

In each case a moving pion loses kinetic energy as it ionises atoms in the emulsion and comes to rest (the photographs show only the last few tens

(Continued)

1 Powell, C. F., *Reports on Progress in Physics* **13**, 1950, 350–424.

Figure 3.1 (Continued)

of microns of the track of these particles). It then decays to give an energetic muon (μ^+). The moving muon loses kinetic energy by ionisation and travels close to $596\,\mu$m before coming to rest. It then decays to give an energetic positron (e^+). Note that the lengths of the muon tracks are very nearly identical in the four cases. This is not an effect of the choice of events to use in the photograph but was true for all positively charged pion decays at rest in the photographic emulsion. The conclusion was that the muons emitted in the decay of pions at rest always have the same kinetic energy and therefore that the pion decay gives a two-body final state. The decay sequence proposed and later verified was

$$\pi^+ \to \mu^+ \to \nu,$$

followed by

$$\mu^+ \to e^+ + \nu + \bar{\nu}.$$

In the first step the muon is produced with a kinetic energy of 4.1 MeV and the unobserved neutrino (ν) has an energy of 29.8 MeV. The three-body decay (including an anti-neutrino $\bar{\nu}$) of the muon means that the positron does not have a unique energy. The pion and muon tracks are deflected from straight by the effect of very many small angle scattering events in the electric field of atomic nuclei. The neutrinos, being neutral, do not ionise, lose no energy, and leave no trace in the photographs.

*3.4   Three-body decays

Consider now the three-body decay

$$a \to 1 + 2 + 3.$$

The notation for the various quantities in the rest frame of 'a' are given in the following table:

	a	→	1	+	2	+	3
Mass	M_a		M_1		M_2		M_3
Momentum	$\mathbf{P}_a = 0$		$\mathbf{P}_1$		$\mathbf{P}_2$		$\mathbf{P}_3$
Energy	$E_a = M_a c^2$		E_1		E_2		E_3

Energy conservation allows the decay if

$$M_a > M_1 + M_2 + M_3.$$

In the rest frame of 'a', conservation of momentum requires

$$\mathbf{P}_a = 0 = \mathbf{P}_1 + \mathbf{P}_2 + \mathbf{P}_3, \tag{3.8}$$

and conservation of energy requires

$$M_a c^2 = E_1 + E_2 + E_3. \tag{3.9}$$

Since this is a three-body final state there is a continuous but finite range of possibilities in the division of the available energy among these particles. However, we can draw some conclusions about the final state of the three particles and on the limits of energy division. These are interesting to deduce since they illustrate the way of thinking about the kinematics. And remember all conclusions are about what would be observed in the rest frame of 'a'.

1   Equation (3.8) tells us that the three momenta are coplanar. (Three vectors which sum to zero may be represented by the three sides of a triangle. A triangle has the property that the three sides are coplanar.)

2   What is the least energy particle 3 can have (call it E_{3L})? Clearly it can be produced at rest in which case $E_{3L} = M_3 c^2$. Then particles 1 and 2 share energy $(M_a - M_3)c^2$ with equal and opposite momenta. This is the same situation as in the two-body decay considered in the last section. The same results apply with the energy available to particles 1 and 2 changed from $M_a c^2$ to $(M_a - M_3)c^2$. Thus, using equation (3.5):

$$E_1 = \frac{(M_a c^2 - M_3 c^2)^2 + (M_1 c^2)^2 - (M_2 c^2)^2}{2(M_a c^2 - M_3 c^2)}$$

and

$$E_2 = \frac{(M_a c^2 - M_3 c^2)^2 + (M_2 c^2)^2 - (M_1 c^2)^2}{2(M_a c^2 - M_3 c^2)},$$

if and only if

$$E_3 = M_3 c^2$$

and, of course, particles 1 and 2 depart with equal and opposite momentum leaving particle 3 at rest.

3   What is the greatest energy particle 3 can have (E_{3G})? Particle 3 will be recoiling against a combination of 1 and 2. Look at equation (3.5) for the final energies in the two-body decay: E_1 is greatest when the mass of the particle recoiling from 1 (in that case particle 2) is least. Similarly, in three-body decay, the energy E_3 is greatest when the mass of the recoiling system is least. That will be when the $1 + 2$ system has no internal kinetic energy and therefore a mass no greater than $M_1 + M_2$. Thus applying equation (3.5):

$$E_{3G} = \frac{(M_a c^2)^2 + (M_3 c^2)^2 - (M_1 c^2 + M_2 c^2)^2}{2M_a c^2}.$$

The final state consists of particles 1 and 2 moving with no relative motion; that means with the same velocity (v). Then we have

$$E_1 + E_2 = M_a c^2 - E_{3G}.$$

But

$$E_1 + E_2 = \frac{(M_1 c^2 + M_2 c^2)}{\sqrt{1 - v^2/c^2}} = E_1(1 + M_2/M_1).$$

Therefore

$$E_1 = \frac{M_1(M_a c^2 - E_{3G})}{(M_1 + M_2)}.$$

E_2 may be found by interchanging subscripts 1 and 2.

4   There is a question that naturally follows from the previous questions. If $E_{3L} < E_3 < E_{3G}$, what are the allowed energy ranges for E_1 and E_2? Finding the answer requires the use of the Lorentz transformations. See Problems 6.18 and 6.19.

Apart from 4, all that was easy. Try Problem 3.2.

*3.5   Two-stage three-body decay

We consider the decay

$$a \rightarrow 1 + 2$$

followed by

$$\downarrow$$
$$2 \rightarrow 3 + 4.$$

The reference frame (Section 3.1) in which 'a' is at rest we label Σ. In Σ, the energies of 1 and of 2 are given by equation (3.5) with appropriate cycling of the labels. In the rest frame of 2, labelled Σ', the energies of 3 and 4 may be found using the same equation. And of course the momenta of 3 and 4 are equal and opposite in Σ'. Because 2 is recoiling in the rest frame of 'a', Σ' is moving in Σ and the energies and momenta of 3 and 4 observed in Σ will depend on the direction of the momenta in Σ' with respect to the direction of motion of 2. The result is to throw 3 and 4 forward with respect to that direction (see Figure 3.2). To find these energies and momenta means performing a Lorentz transformation from Σ' to Σ. We shall show how to do this in Section 6.2.

There are many examples of this situation in elementary particle physics. In some cases the average lifetime of 2 is sufficiently long that it moves a detectable distance in the laboratory; we shall examine such an example in Section 3.6. In many cases the average lifetime of 2 can be so short that it does not move an experimentally observable distance before decaying and its presence must be deduced indirectly. Here is an example. There is a D^+ meson and one of its decay modes is observed to be

$$D^+ \rightarrow \pi^+ + \pi^+ + K^-$$

Masses in MeV/c^2:   1869   139.6   139.6   493.6

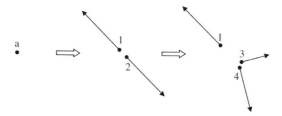

Figure 3.2 The two-stage three-body decay a → 1 + 2 → 1 + 3 + 4. In the rest frame of 2 its decay products 3 and 4 are emitted with equal and opposite momenta. The motion of 2 before its decay throws these decay products forward in the direction of its motion. This is an example of the difference between rest frames and frames in which particles are moving. See Figure 3.4 for two examples.

These three decay products are familiar and although they do decay they have average lifetimes long enough to allow identification and a measurement of momenta. However, in the rest frame of the D$^+$, one of the π^+ mesons frequently has an energy close to 727 MeV (Figure 3.3). This means that it is the product of a two-body decay and that in such cases the complete decay occurs in two steps:

Step 1:   D$^+ \rightarrow \pi^+ +$ X
                    $\downarrow$
Step 2:   X $\rightarrow \pi^+ +$ K$^-$.

We can calculate the mass of X as follows. Applied here, equation (3.5) reads

$$E_\pi = \frac{(M_D c^2)^2 + (M_\pi c^2)^2 - (M_X c^2)^2}{2(M_D c^2)}$$

with obvious notation. Everything in this equation is known except the M_X to be calculated. Hence

$$M_X c^2 = \sqrt{(M_D c^2)^2 + (M_\pi c^2)^2 - 2(M_D c^2)E_\pi}.$$

The result is $M_X c^2 = 892$ MeV. This particle X is given the symbol $\overline{\text{K}}^{*\circ}$. (The neutral anti-K*!)

There is another way of obtaining the mass of particle X. This depends on the fact that in many circumstances the momentum of both decay products may be measured. This method is discussed in Section 3.6.

The Heisenberg uncertainty relation implies that the finite average lifetime of an unstable particle gives a spread of the individual particle mass values around the mean value. In this case the $\overline{\text{K}}^{*\circ}$ mesons have an average lifetime ($\cong 1.3 \times 10^{-23}$ s) short enough for the mass uncertainty (± 25 MeV/c^2) to have a detectable effect on the kinematics. Decay by decay the 727 MeV energy of the first π^+ will vary by plus or minus an amount that has an average over many such decays of about 12 MeV.

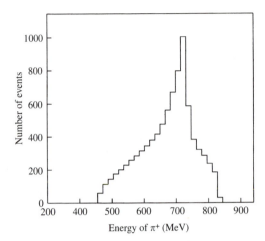

Figure 3.3 The energy spectrum of the most energetic positive pion emitted in a large sample of events each containing the decay

$$D^+ \rightarrow \pi^+ + \pi^+ + K^-.$$

This energy is that measured in the rest frame of the D^+. The peak at 730 MeV corresponds to a large contribution from two stage decays in which one pion is recoiling against a short-lived particle which later decays into a π^+ and a K^-. The peak sits on a broad spectrum of π^+ energies corresponding to a direct decay to the same three-body final state. (Data freely adapted from Frabetti *et al.*, *Physics Letters* **B331**, 1994, 217–26.)

Such a short average lifetime means this particle is one that will not move a detectable distance before decay. In this decay of the D^+, the $\overline{K}^{*0}$ will, on average, move from production to its own decay a distance of about 9×10^{-15} m in the rest frame of the D^+.

D^+ mesons are normally produced in certain high energy particle collisions and are not at rest. Their average life is about 10^{-12} s. Their mass uncertainty is negligibly small compared to their mass. The distance moved between production and decay can be of the order of 1 mm. Measuring that kind of distance in what is a challenging production environment can be done with modern instrumentation.

**3.6 The decay of a moving particle

Consider the decay

$$a \rightarrow 1 + 2 + 3 + \cdots.$$

We apply conservation of energy and momentum to the case in which 'a' is moving. With obvious notation we have:

$$E_a = E_1 + E_2 + E_3 + \cdots = \sum_i E_i$$

and

$$\mathbf{P}_a = \mathbf{P}_1 + \mathbf{P}_2 + \mathbf{P}_2 + \cdots = \sum_i \mathbf{P}_i, \quad i = 1, 2, 3, \ldots .$$

Also

$$E_i = (P_i c)^2 + (M_i c^2)^2, \quad i = a, 1, 2, 3, \ldots .$$

In particular we have for particle 'a' using equation (3.1)

$$
\begin{aligned}
(M_a c^2)^2 &= E_a^2 - \mathbf{P}_a \cdot \mathbf{P}_a c^2 \\
&= \left(\sum_i E_i\right)^2 - \left(\sum_i \mathbf{P}_i\right) \cdot \left(\sum_i \mathbf{P}_i\right) c^2, \quad i = 1, 2, 3, \ldots . \quad (3.10)
\end{aligned}
$$

Here we have extended our use of vector notation. See Table 3.1.

Thus we can determine the mass of the decaying particle if we can measure the energy and momentum of all the decay products. Normally this becomes the problem of measuring the momenta or energy and identifying the products. Experimentally, the possibility of missing one or more products, by reason of inefficiencies or incomplete detection cover, must be kept in mind.

A very striking example of the application of this technique is provided by Figure 3.4. The means of obtaining this photograph are described in the caption. It shows the trajectories of the fast charged particles in liquid hydrogen with an imposed magnetic field. There is a lot of information in, or associated with, this photograph and its stereo partners:

1   The particles incident from the left are negative pions (π^-) of momentum 1000 MeV/c.
2   One π^- collides with a proton (p) of the hydrogen. The proton is essentially at rest.
3   The momentum vectors, charges and, with considerable confidence, the identity of the particles labelled 1 to 4 could be determined. (π^-, π^+, p, π^-)

The conclusions that could be drawn were

1   Momentum and energy were conserved overall from initial $\pi^- + p$ to particles 1 through 4.
2   Hence no other particles were produced in this sequence of events and evaded detection.

We label by V_0 the point where the incident pion seems to disappear (the primary vertex in the jargon of the subject) and by V_1 and V_2 the secondary vertices where the final state charged particles appear. Then, additional conclusions may be drawn:

3   There was a single neutral particle k that moved from V_0 to V_1 where it decayed into a π^- and a π^+. The energies and momenta of these two pions gave the mass of k by the use of the two-body form of equation (3.10) ($i = 1, 2$).

4   Similarly there was a single neutral particle λ that moved from V_0 to V_2 where it decayed into a π^- and a proton (p). As in 3 above the mass of λ could be found.

5   The results of such measurements were consistent with the now well-known values for these particles. Their existence has been confirmed in many events, in many experiments.

The k is now called the K^0 (neutral kaon) and has a mass of $497.7\,\mathrm{MeV}/c^2$.

The λ is now called the Λ (lambda baryon) and has a mass of $1115.7\,\mathrm{MeV}/c^2$.

Note that although these two masses could be found by measurements on their decay products, other information from the event can be used to improve the precision of the results. Suppose the following are given:

- energy and momentum conservation at all vertices,
- the energy of the incident pions and the applied magnetic field, with errors,
- measurements of the true space positions, with errors, of the vertices and of many points along each visible trajectory,
- the hypothesis that the reaction is

$$\pi^- + p \to k + \lambda$$
$$\begin{array}{l} \quad \longrightarrow \lambda \to \pi^- + p \\ \quad \longrightarrow k \to \pi^- + \pi^+. \end{array}$$

Then a statistical method may be used: The masses of the k and of the λ and of other kinematic unknowns are varied until a reconstruction of the event is found that best fits all the actual measurements and their errors. In this way all measurements on the event contribute to maximise the precision on the masses and on the kinematic reconstruction.

The decays at V_1 and V_2 are two-body decays as discussed in Section 3.3. Therefore in a reference frame in which either parent k, or λ, is at rest, the decay products have opposite momenta. The motion of the parent throws the momenta forward with the consequence that in each case the angle in the laboratory between the decay momenta is less than $180°$. This is the same effect as mentioned at the beginning of Section 3.5. Figure 3.4 gives two clear examples of this.

Try Problem 3.3.

3.7   Kinematics of formation[†]

Suppose we have a particle 1 that decays:

$$1 \to a + b.$$

[†] Note that this is the first section to be found without * or ** against the header. This indicates material that may be omitted from a first course in special relativity.

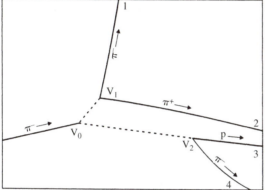

Figure 3.4 A bubble chamber photograph of the event described in Section 3.6. A
negative pion (π^-) with a momentum of 1 GeV/c collides with a stationary
proton at V_0. Two energetic, electrically neutral particles are produced. One
decays at V_1 to produce a π^- and a π^+. The other decays at V_2 to produce
a proton (p) and a π^-.

The bubble chamber consists of a body of liquid, in this case liquid
hydrogen at a temperature near 20 K, under pressure. When the pressure
is released the liquid becomes super-heated and bubbles begin to form.
Initially these bubbles grow on the trail of ions left by the passage of fast
charged particles. By correct timing of the accelerator that provides the
incident particles, of the pressure release, and of the flash light, photographs
of this kind may be obtained. In this photograph the charged particles
are losing kinetic energy as they cause repeated ionisation. However, the
fractional loss of energy is small (except for particle 4). In the case of the
muon tracks in Figure 3.1, the energy loss is sufficient to bring each muon
to rest.

The reverse of this decay is possible and is called **formation**. Thus, if the kinematic conditions are correct it is possible to *form* particle 1 in a collision: once formed 1 will later decay.

$$a + b \rightarrow 1 \rightarrow a + b.$$

This may not be the only decay mode allowed by energy conservation. If it happens that there is no other decay mode, we know that this one must exist because it is just the reverse of the energetically allowed formation that produced particle 1. Now we consider an experimental situation in which 'b' is at rest in a reference frame that is stationary in the laboratory (*the lab frame*, in the jargon of the trade) which we signify by Σ_{lab}. Particle 'a' is incident with energy E_a. What value E_a allows the formation of 1? Let us tabulate our notation.

Collision in the Lab Frame Σ_{lab} ('b' at rest)

	a	+	b	→	1
Mass	M_a		M_b		M_1
Momentum	P_a		$P_b = 0$		P_1
Energy	$E_a = \sqrt{(P_a c)^2 + (M_a c^2)^2}$		$E_b = M_b c^2$		E_1

Conservation of momentum:

$$\mathbf{P}_1 = \mathbf{P}_a.$$

Conservation of energy:

$$E_1 = E_a + M_b c^2.$$

But we require

$$E_1^2 - (P_1 c)^2 = (M_1 c^2)^2$$

which in this case means

$$(E_a + M_b c^2)^2 - (P_a c)^2 = (M_1 c^2)^2.$$

From this equation the value of E_a that allows the formation of '1' is given by

$$E_a = \frac{(M_1 c^2)^2 - (M_a c^2)^2 - (M_b c^2)^2}{2(M_b c^2)}. \qquad (3.11)$$

Once formed, particle 1 has a fractional speed in the laboratory given by

$$\beta_1 = P_1 c / E_1 = P_a c / (E_a + M_b c^2). \qquad (3.12)$$

As indicated, once formed particle 1 will decay. If it is still moving at decay, once again the momenta of the decay products in the rest frame of 1 are thrown forward by its motion. This is the same effect as we have met in Sections 3.5 and 3.6.

3.8 Examples of formation

How does formation show up? In a + b → 1 → ··· the presence of state 1 can greatly enhance the probability of the scattering or absorption of particle 'a' incident on a target of particles 'b' when 'a' has the correct energy. Experimentally this enhancement may show up in a plot of the total cross-section for a collision of 'a' with 'b'. There can be a strong peak at the energy for formation. Such peaks are sometimes called resonances.

1   A non-relativistic but a very striking example: Figure 3.5 shows the absorption cross-section for slow neutrons in indium as a function of the neutron kinetic energy. Near 1 eV, there is a huge peak due to the formation of an excited state of an isotope of indium:

$$n + {}^{115}_{49}\mathrm{In} \rightarrow {}^{116}_{49}\mathrm{In}^*,$$

$$a + b \rightarrow 1.$$

This excited state (${}^{116}_{49}\mathrm{In}^*$) has only a small probability of decaying by the reverse of the formation. Most decays are by the emission of successive photons to the ground state of ${}^{116}_{49}\mathrm{In}$. The finite lifetime of the excited state means that its mass is not unique but uncertain, as for the $\overline{\mathrm{K}}^{*\circ}$ in Section 3.5.

2   An example of relativistic formation: Figure 3.6 shows the total cross-section for the scattering of positive pions by protons (stationary) as a function of the momentum (in the laboratory) of the incident pions. There is a very

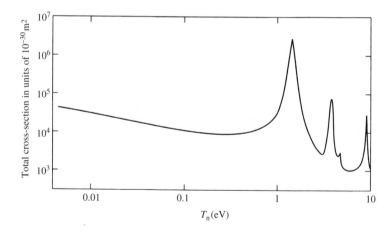

Figure 3.5 The total cross-section for neutrons in natural indium (96% ${}^{115}_{49}\mathrm{In}$) as a function of the neutron kinetic energy T_n in the range 5 meV to 10 eV. Note that both scales are logarithmic so that, in particular, the prominence of the peaks above the continuum is greatly reduced in this figure. On a linear scale the peak near 1 eV is nearly one hundred times greater than the neighbouring peaks.

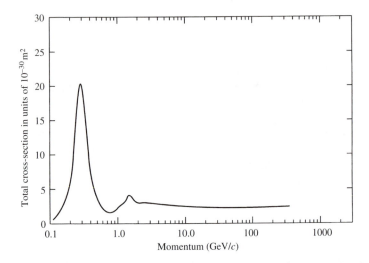

Figure 3.6 The total cross-section for the scattering of positive pions by protons as a function of the momentum in the laboratory of the incident mesons. The momentum scale is logarithmic but the cross-section scale is linear. The peak at 300 MeV/c is due to the formation of an unstable baryon (a relative to the proton) labelled Δ^{++}.

striking maximum in the cross-section at a momentum of about 300 MeV/c. The interpretation is that the reaction is dominated by a formation and decay of an unstable particle of mass 1232 MeV/c^2:

$$\pi^+ + p \rightarrow \Delta^{++} \rightarrow \pi^+ + p.$$

Masses in MeV/c^2:   139.6   938.3   1232

a   + b → 1.

We use equation (3.11) to calculate the incident π^+ laboratory energy at the maximum:

$$E_\pi = \frac{(1232)^2 - (139.6)^2 - (938.3)^2}{2 \times 938.3}$$

$$= 329 \text{ MeV}.$$

The momentum of the incident π^+ at maximum is given by

$$P_\pi c = \sqrt{E_\pi^2 - (M_\pi c^2)^2}$$

$$= 298 \text{ MeV}.$$

Again the mass of the particle formed is uncertain. This is a consequence of its very short mean life, about 6×10^{-24} s. The width of the peak in the cross-section corresponds to a mass distribution that has a full width at half maximum that is about $120 \, \text{MeV}/c^2$.

Try Problem 3.4.

**3.9 Photons

In Section 2.3 we found that the equation (2.8) $E = Pc$ was correct for photons. This implies that kinematically they behave as do particles with zero mass. None of the applications we have discussed has involved any photons. The reason is simple: the photon does not require special treatment. If any photons are involved every formula is still correct and it is only necessary to put every photon's mass equal to zero. In fact we have already done something similar in the example of the charged pion decay, Section 3.3, where a neutrino is emitted; this particle either has zero mass or a mass negligibly small in its effect on the kinematics of the decay. We put $M_\nu = 0$, correctly. Do the same for any photon. Try Problems 3.5 and 3.6.

Note that neither a photon nor any zero mass particle has a rest frame. All such objects move with the speed of light and it is impossible to move a reference frame, with its observer, at that speed.

**3.10 Compton scattering

In 1923 A. H. Compton investigated the decrease in penetrating power of X-rays from before to after being scattered by targets of light elements. He found that one component of the scattered radiation had a wavelength greater than that of the incident X-rays. He attributed this change as being due to the elastic scattering of photons (γ) by nearly free electrons:

$$\gamma + e^- \rightarrow \gamma + e^-.$$

He found agreement between his measured wavelength changes and those calculated using relativistic kinematics. This agreement was important for two reasons:

i   It confirmed the quantum nature of X-rays.
ii   It confirmed the correctness of relativistic particle kinematics in what was its first application to an observed process.

There are some points to be made about this process:

1   Elastic scattering means that it is a two-body to two-body process and neither particle changes its intrinsic nature. In this case there is a photon in the initial state and one photon in the final state. However, in general there will be a change in wavelength. And the electron remains an electron.
2   Free electrons do not exist in sufficient number and density for this process to be observed in a terrestrial experiment under conditions free of complications.

3   The laboratory observation of the process will consist of a beam of mono-energetic photons directed at a suitable stationary target with arrangements to measure the wavelength of any scattered photons. This is a *fixed target* experiment (see Section 3.11). The 'free' electron is provided by the electrons bound in the target material. The electron will recoil from any collision and the photon loses energy. Thus there will be a connection between the energy of the scattered photon and the angle through which it has scattered. If the photon energy is large compared with the binding energy of electrons in the target material then the effects of the bound state of the struck electron on this connection will be small. Compton used, in one of a series of measurements, 17.5 keV X-rays and carbon targets.

The relativistic kinematic analysis of Compton scattering can be done with the tools available to us at this stage. We shall do the analysis of the fixed target experiment, that is with the initial state electron at rest in the laboratory. Figure 3.7 shows the before and after momenta. We define our notation in a table:

Collision quantities in the laboratory

	γ +	e^- $\rightarrow$	γ +	e^-
Label	1	2	3	4
Mass	0	M_e	0	M_e
Momentum	$\mathbf{P}_1$	$\mathbf{P}_2 = 0$	$\mathbf{P}_3$	$\mathbf{P}_4$
Energy	$E_1 = P_1 c$	$E_2 = M_e c^2$	$E_3 = P_3 c$	$E_4 = \sqrt{(P_4 c)^2 + (M_e c^2)^2}$

Apply conservation of energy and momentum:

$$E_1 + M_e c^2 = E_3 + E_4,$$
$$\mathbf{P}_1 = \mathbf{P}_3 + \mathbf{P}_4.$$

We want to find the magnitude and direction (angle θ between $\mathbf{P}_3$ and $\mathbf{P}_1$) of the momentum of the scattered photon, $\mathbf{P}_3$. This means eliminating E_4 and $\mathbf{P}_4$ from

Figure 3.7 Compton scattering in the laboratory. A photon of momentum $\mathbf{P}_1$ strikes a stationary electron and scatters at an angle θ with momentum $\mathbf{P}_3$. The electron recoils at angle ϕ and a momentum $\mathbf{P}_4$. Conservation of momentum means that $\mathbf{P}_4$ lies in the plane containing $\mathbf{P}_1$ and $\mathbf{P}_3$.

these equations. To do this form

$$(M_e c^2)^2 = E_4^2 - (\mathbf{P}_4 c) \cdot (\mathbf{P}_4 c)$$
$$= (E_1 + M_e c^2 - E_3)^2 - (\mathbf{P}_1 - \mathbf{P}_3) \cdot (\mathbf{P}_1 - \mathbf{P}_3) c^2.$$

Collecting terms from left and right and using $(\mathbf{P}_1 \cdot \mathbf{P}_1) c^2 = (P_1 c)^2 = E_1^2$ and $(\mathbf{P}_2 \cdot \mathbf{P}_2) c^2 = (P_2 c)^2 = E_2^2$ for the initial and final photons, and $\mathbf{P}_1 \cdot \mathbf{P}_3 = P_1 P_3 \cos \theta$ gives

$$\frac{1}{E_3} - \frac{1}{E_1} = \frac{1}{M_e c^2}(1 - \cos \theta). \tag{3.13}$$

For a photon $\lambda = hc/E$ where h is Planck's constant, so we obtain

$$\lambda_3 - \lambda_1 = \frac{h}{M_e c}(1 - \cos \theta), \tag{3.14}$$

where the quantity $h/M_e c$ has the dimensions of length and is called the Compton wavelength of the electron. Its value is 2.426×10^{-12} m.

Equation (3.13) is rederived in Section 9.2 as an illustration of the use of four-vectors.

Equation (3.14) clearly expresses the change in wavelength that Compton measured. One of his results is shown in Figure 3.8. In this case he measured a wavelength shift of 2.36×10^{-12} m where 2.43×10^{-12} m was expected. He also found the expected dependence of the wavelength shift on the angle of scattering and shifts independent of target material. His work is considered a milestone in establishing the quantum nature of the entire spectrum of electromagnetic radiation.

The Compton scattering of photons from a laser beam by a beam of energetic electrons has been used to produce a beam of high energy, mono-energetic photons for research purposes (Problem 6.17). A similar arrangement has been used to measure beam properties at the LEP collider at The European Centre for Nuclear Research (CERN) (Section 4.13).

**3.11   The centre-of-mass

Section 3.6 should have prepared the reader for the next step in which we consider a set of particles S in a certain reference frame Σ.

Masses: $M_1, M_2, M_3, \ldots$.
Momenta: $\mathbf{P}_1, \mathbf{P}_2, \mathbf{P}_3, \ldots$.
Energies: $E_1, E_2, E_3, \ldots$.

The total momentum is

$$\mathbf{P} = \mathbf{P}_1 + \mathbf{P}_2 + \mathbf{P}_3 + \cdots = \sum_i \mathbf{P}_i \tag{3.15}$$

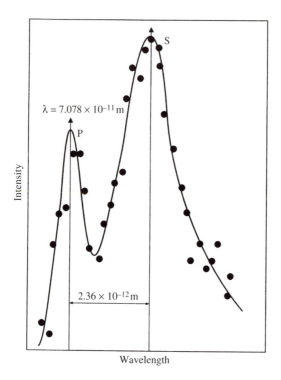

$\lambda = 7.078 \times 10^{-11}\,\text{m}$

P

S

Intensity

$2.36 \times 10^{-12}\,\text{m}$

Wavelength

Figure 3.8 The wavelength spectrum of scattered X-rays observed at an angle of 90°
to an incident beam of X-rays of wavelength 7.078×10^{-11} m (17.52 keV)
incident on a carbon target. The peak P has the same wavelength as the
incident radiation and is due to elastic scattering of X-ray photons by
carbon atoms. The peak S is at a wavelength longer by 2.36×10^{-12} m
and is due to the scattering of photons by lightly bound electrons in carbon
atoms. This figure is a modified copy from one of Compton's original
papers. The expected wavelength shift for scattering from free electrons is
2.43×10^{-12} m.

and the total energy is

$$E = E_1 + E_2 + E_3 + \cdots = \sum_i E_i. \tag{3.16}$$

Note that we are using *total energy* to mean the sum over all particles of what,
originally, for each particle, we called the total energy, but which we now call the
energy (Section 2.2).

There are reference frames in which the total momentum is zero. They differ
only in the orientation of the axes and we can choose one according to the needs
of the problem in hand. So the **centre-of-mass** is this chosen frame and is labelled
Σ_{cm}. We add a prime to indicate the energies and momenta of individual particles

in Σ_{cm}. Thus

$$\mathbf{P}' = \sum_i \mathbf{P}'_i = 0, \quad i = 1, 2, 3, \ldots,$$

and

$$E' = \sum_i E'_i, \quad i = 1, 2, 3, \ldots \equiv W.$$

Now E' is the total of all the energies in the reference frame Σ_{cm}. As such we give it the distinctive symbol W. It is called the **centre-of-mass energy** although more strictly it should be called the *energy in the centre-of-mass*. The problem with this distinction will become clear shortly.

The quantity $\sqrt{E_1{}^2 - \mathbf{P}_1 \cdot \mathbf{P}_1 c^2}$ for single particle 1 is its rest mass energy $M_1 c^2$ or, more explicitly, its energy in a reference frame in which its momentum is zero. Note that this is true in whatever reference frame E_1 and $\mathbf{P}_1$ are measured. For a group of particles, $\sqrt{E^2 - \mathbf{P} \cdot \mathbf{P} c^2}$ formed from $\mathbf{P}$ and E of equations (3.15) and (3.16) and measured in any frame, is the energy of all the particles in the frame in which the group as a whole has no total momentum. That reference frame is the centre-of-mass and the energy is W. Thus

$$\sqrt{E^2 - \mathbf{P} \cdot \mathbf{P} c^2} = \sqrt{E'^2 - \mathbf{P}' \cdot \mathbf{P}' c^2} = E' \equiv W, \tag{3.17}$$

and hence, from equations (3.15) and (3.16)

$$W^2 = \left(\sum_i E_i \right)^2 - \left(\sum_i \mathbf{P}_i c \right)^2, \quad i = 1, 2, 3, \ldots. \tag{3.18}$$

Therefore the centre-of-mass energy W for a set of particles may be found if all their energies and momenta are known in any reference frame. (We have put $\left(\sum_i \mathbf{P}_i \right) \cdot \left(\sum_i \mathbf{P}_i \right) = \left(\sum_i \mathbf{P}_i \right)^2$.)

Returning to the frame Σ we have the group of particles S. The velocity of the centre-of-mass (and of the frame in which it is stationary, Σ_{cm}) in Σ may be found using the momentum $\mathbf{P}$ and energy E of this group of particles in equation (3.2). We find

$$\mathbf{v}_{cm} = \mathbf{P} c^2 / E, \tag{3.19}$$

and its Lorentz factor is

$$\gamma_{cm} = E / W, \tag{3.20}$$

where $\mathbf{P} c$ and E are given by equations (3.15) and (3.16).

It is worth recalling that we are discussing a system of particles S. For simplicity we shall only be considering the cases in which the particles in S are all the products of a collision or of a decay, or in which a two-particle system is moving towards a collision. In all cases the energies and momenta of all the particles are those they have when they are free, that is when they are not under the influence of the interactions that cause the collisions or decays.

One of the difficulties with the concept of the centre-of-mass in relativistic kinematics is a slight confusion of terminology caused by usage. In classical mechanics a rigid body will have a centre-of-mass, sometimes loosely called the centre-of-gravity. It is a point fixed with respect to the body. A system of stationary particles will also have a centre-of-mass. If that situation is an instantaneous photograph of particles that are capable of relative motion and are actually moving then this centre-of-mass may also be moving. A frame in which the vector sum of the momenta of the particles is zero is one in which this centre-of-mass is at rest. Strictly it should be called the *centre-of-momentum*. But usage has fixed on *centre-of-mass*. There is another difficulty. For a single particle we might speak of the particle's energy in a frame Σ. Its centre-of-mass is a frame in which the particle is at rest. The energy in that frame is the rest mass energy. The phrase *centre-of-mass energy* might be thought to mean the energy of the moving centre-of-mass in some specified frame, Σ say. That is E in equation (3.15). It is not! Usage has made it the energy in the centre-of-mass frame, that is W of equation (3.18).

Summarising:

> W = the *centre-of-mass energy*,
> = the sum of the energies of all the particles of S
> in their centre-of-mass frame,
> = the rest mass energy of the particle that if it existed might
> decay into the set of particles S. (W/c^2 is sometimes called
> the *invariant mass*.)

Why is the centre-of-mass so important in the context of this chapter? To answer this question consider two-body collisions in the laboratory. In nuclear and particle physics that has often come to mean a beam of energetic particles (a) striking a stationary target of other particles (b). Sometimes this is called a *fixed target* experiment. Suppose that in some of these collisions there are nuclear or particle reactions such that the products are different from the colliding particles. In particular, if we wish to create new particles in a collision we must convert some kinetic energy into mass. In the centre-of-mass the absolute limit to this conversion occurs if the sum of the rest mass energies of the final state particles reaches W. At this limit these particles would be at rest in the centre-of-mass frame and there is no kinetic energy left to create others. The argument about this limit is based on what is happening in the centre-of-mass; but by the Principle of Relativity, what cannot happen in the centre-of-mass frame cannot happen in any reference frame. Thus the usefulness with respect to available energy of any collision is limited by the value of the centre-of-mass energy, W. In any other frame, including the laboratory frame of a fixed target experiment, some of the incident kinetic energy is tied up in the conserved momentum of the motion of the centre-of-mass system as a whole.

To end this section it is worth adding a list of the equivalences that exist between kinematic quantities for a single moving particle and for the moving centre-of-mass of a group of particles. See Table 3.2. Thus, for example, a given Lorentz transformation (Chapter 6) will transform the energy and momentum of the

Table 3.2 Equivalent kinematic quantities for a particle and a centre-of-mass system as observed from a reference frame in which neither is at rest

Moving particle		Moving centre-of-mass system	
Name of quantity	*Symbol*	*Name of quantity*	*Symbol**
Rest mass energy	Mc^2	Centre-of-mass energy	W
Energy	E	Energy	$\sum_i E_i$
Momentum	$\mathbf{P}$	Momentum	$\sum_i \mathbf{P}_i$
Fractional velocity	$\beta = \mathbf{P}c/E$	Fractional velocity	$\beta_{cm} = \sum_i \mathbf{P}_i c / \sum_i E_i$
Lorentz factor	$\gamma = E/Mc^2$	Lorentz factor	$\gamma_{cm} = \sum_i E_i / W$
	$(Mc^2)^2 = E^2 - \mathbf{P} \cdot \mathbf{P}c^2$		$W^2 = \left(\sum_i E_i\right)^2 - \left(\sum_i \mathbf{P}_i\right) \cdot \left(\sum_i \mathbf{P}_i\right) c^2$

* $\sum_i$ here means the sum over all the particles constituting the system.

centre-of-mass $(\sum E_i, \sum \mathbf{P}_i c)$ in the same way as it does the energy and momentum of a single particle $(E, \mathbf{P}c)$.

Again we anticipate that we have to transform energies and momenta between reference frames. In the context of this section that could be between Σ_{cm} and Σ. (See Sections 6.8 and 6.9.)

**3.12 The centre-of-mass in two-body collisions

In this section we extend our discussion of collisions in the case of fixed target experiments. A beam of energetic particles (a) strikes a target of other particles (b) which are at rest. The table shows our notation. It looks similar to the one displayed in Section 3.7.

Collision in the laboratory frame Σ_{lab} ('b' at rest)

	a	+	b	→	centre-of-mass
Mass	M_a		M_b		$W =$ energy in centre-of-mass
Momentum	$\mathbf{P}_a$		$\mathbf{P}_b = 0$		$\mathbf{P} =$ momentum of centre-of-mass in $\Sigma_{lab} = \mathbf{P}_a$
Energy	$E_a = \sqrt{(P_a c)^2 + (M_a c^2)^2}$		$E_b = M_b c^2$		$E =$ energy (total) of centre-of-mass in Σ_{lab} $= E_a + M_b c^2$

We can now find W easily using equation (3.18):

$$W^2 = E^2 - (Pc)^2 = (E_a + M_b c^2)^2 - (P_a c)^2$$
$$= (M_a c^2)^2 + (M_b c^2)^2 + 2E_a(M_b c^2). \qquad (3.21)$$

The velocity of the centre-of-mass in Σ_{lab} is found using equation (3.19)

$$\mathbf{v}_{\text{cm}} = \mathbf{P}c^2/E = \mathbf{P}_a c^2/[E_a + M_b c^2] \qquad (3.22)$$

and the Lorentz factor from equation (3.20)

$$\gamma_{\text{cm}} = E/W = (E_a + M_b c^2)/W. \qquad (3.23)$$

If the incident energy is so great that $E_a \gg M_a c^2$ and $M_b c^2$ (the extreme relativistic situation) then

$$W = \sqrt{2E_a M_b c^2}. \qquad (3.24)$$

In the next section we examine an application of equation (3.24).

In Section 3.7, equation (3.11), we found the condition required on the collision of $a + b$ ('b' at rest) in order that the outcome can be the formation of the state 1 of mass M_1. In the context of the idea of the centre-of-mass this condition is the same as requiring $W = M_1 c^2$.

*3.13   Reaction thresholds in the laboratory

Consider the production reaction we saw in Figure 3.4:

	π^-	$+$	p	$\rightarrow$	K^0	$+$	Λ
	M_π		M_p		M_K		M_Λ
Masses in MeV/c^2:	139.6		938.3		497.7		1115.7

with the target proton (p) at rest in the laboratory reference frame Σ_{lab}. Since

$$M_\pi + M_p < M_K + M_\Lambda$$

we have to convert some incident kinetic energy into rest mass. As the momentum of the incident pion increases from zero, W increases from $(M_\pi + M_p)c^2$. When W reaches $(M_K + M_\Lambda)c^2$ the reaction becomes possible. This is the **threshold** for the reaction. The corresponding E_π is the **reaction threshold energy**.

Generalising to a many body final state (but 'b' still at rest in the laboratory)

$$a + b \rightarrow 1 + 2 + 3 + \cdots.$$

At threshold (subscript th)

$$W_{\text{th}} = M_1 c^2 + M_2 c^2 + M_3 c^2 + \cdots.$$

Then, remembering 'b' is at rest we have from equation (3.21)

$$(M_ac^2)^2 + (M_bc^2)^2 + 2E_{a,th}(M_bc^2) = (M_1c^2 + M_2c^2 + M_3c^2 + \cdots)^2.$$

This gives for the energy $E_{a,th}$ in the laboratory of the incident particle 'a' at threshold:

$$E_{a,th} = \frac{1}{2M_bc^2}\left\{(M_1c^2 + M_2c^2 + M_3c^2 + \cdots)^2 - (M_ac^2)^2 - (M_bc^2)^2\right\}.$$

$$(3.25)$$

If you prefer the threshold laboratory kinetic energy then that is given by

$$T_{a,th} = E_{a,th} - M_ac^2$$

$$= \frac{1}{2M_bc^2}\left\{(M_1c^2 + M_2c^2 + M_3c^2 + \cdots)^2 - (M_ac^2 + M_bc^2)^2\right\}.$$

$$(3.26)$$

Do Problems 3.7–3.12. These concern various centre-of-mass matters and threshold conditions for some collision reactions.

*3.14   Summary

We have come further in relativistic kinematics than may be required for confidence building alone. We have developed some important techniques and the fact that this has been done using only two formulae, equations (3.1) and (3.2), stresses the simplicity of relativistic kinematics. However, we did note in Sections 3.5–3.7 and in 3.11, that we have to be prepared to change reference frames. That brings us right up against the Lorentz transformations for energy and momentum. We shall explore that matter in Chapter 6. Although none of the mathematics is difficult, there are techniques using four-vectors that can simplify some of this relativistic kinematics. Therefore, we shall return again to the subject in Section 9.2, after we have introduced four-vectors in Section 8.10.

Problems

Use the units MeV, MeV/c, GeV and so on, as appropriate.

3.1   The Λ baryon has several decay modes. One is

$$\begin{array}{ccccc} \Lambda & \rightarrow & \pi^- & + & p \\ 1115.7 & & 139.6 & & 938.3 \end{array}$$

where the masses in MeV/c^2 are the figures below the symbols for each particle involved. Calculate the values in the Λ rest frame for the kinetic energy, momentum, and speed for each decay product.

3.2   The K$^+$ meson has several decay modes one of which is to three charged pions:

$$K^+ \quad \rightarrow \quad \pi^+ \quad + \quad \pi^+ \quad + \quad \pi^-.$$
$$493.7 \qquad 139.6$$

Calculate the maximum kinetic energy in the K-meson rest frame that any pion may have in this decay. (The π^- and the π^+ have the same mass.)

3.3   A neutral particle L^0 is observed in bubble chamber to decay into a negative pion and a proton:

$$L^0 \rightarrow \quad \pi^- \quad + \quad p.$$
$$139.6 \qquad 938.3$$

The momenta of the pion and proton are measured by curvature of their trajectories in a magnetic field to be 250 and 1000 MeV/c respectively. The space angle between the trajectories at the point of decay is 21°. Calculate the mass of the L^0.

3.4   A particle of rest mass M and kinetic energy $3Mc^2$ strikes a stationary particle of rest mass $2M$ and sticks to it. Find the rest mass and speed of the composite particle.

3.5   In Problem 2.7 the reader was asked to find some energies in nuclear emission and absorption of photons. Derive the required formula using fully relativistic kinematics.

3.6   A photon rocket is a device which propels itself by emitting photons, all in the same direction. It has initial rest mass M. Show that when the rest mass has fallen to αM the speed of the rocket as measured in the starting rest frame is given by

$$v = c\frac{1 - \alpha^2}{1 + \alpha^2}.$$

3.7   A charged pion with a momentum of 1 GeV/c collides with a proton stationary in the laboratory. Calculate

a   the centre-of-mass energy,
b   the speed of the centre-of-mass in the laboratory, and
c   the momentum of one of the particles in the centre-of-mass.

3.8   A neutral pion may be produced in an energetic proton–proton collision:

$$p \quad + \quad p \rightarrow p + p + \pi^0.$$
$$938.3 \qquad\qquad\qquad 135.0$$

Calculate the kinetic energy of a proton incident on a stationary proton at the threshold for this reaction. Similarly calculate the threshold kinetic energy of the pion for the reaction

$$\pi^- + p \rightarrow K^0 + \Lambda.$$

This is the production reaction in Figure 3.4. The K^0 rest mass is 493.7 MeV/c^2. The remaining masses are given in Problem 3.1.

3.9 An electron with kinetic energy 10 GeV strikes a proton at rest. What is the speed of the centre-of-mass system? What energy is available to produce new particles in addition to the electron and proton?

3.10 An electron collides with a second electron stationary in the laboratory to produce a pair of muons thus:

$$e^- + e^- \rightarrow e^- + e^- + \mu^+ + \mu^-.$$

Show that in terms of the mass of the electron (M_e) and of the muons (M_μ) this reaction can only occur if the momentum of the incident electron exceeds P_{th} given by

$$P_{th} = 2M_\mu c(1 + M_\mu/M_e)\sqrt{1 + 2M_e/M_\mu}.$$

3.11 Two photons collide to produce an electron–positron pair:

$$\gamma + \gamma \rightarrow e^- + e^+.$$

The two photons have anti-parallel momenta. One has energy E_0, the other energy E. Obtain an expression in terms of E_0 and the electron mass M_e, for the energy E at the threshold for this reaction.

High energy photons of galactic origin pass through the 2.7 K cosmic microwave background radiation which can be regarded as a gas of photons of energy 2.3×10^{-4} eV. Calculate the threshold energy for the production of electron–positron pairs.

3.12 Prove

a   that an electron and a positron in collision cannot annihilate to produce a single photon,

b   that a photon cannot give up all its energy in a collision with a free electron,

c   that an isolated photon cannot give rise to an electron–positron pair, and

d   an isolated electron moving with speed v in a vacuum cannot emit a single photon.

A hydrogen atom in an excited state, moving in a vacuum, can emit a single photon. Why is this different from (d)?

4 Electrodynamics

**4.1 Introduction

What do we mean by **dynamics**? It is the branch of mechanics that is concerned with the forces that change or produce the motion of particles or bodies. It is distinct from *kinematics* which is concerned with the mechanics of particles or bodies without reference to the forces involved. Thus Chapter 3 on relativistic kinematics investigated conservation of energy and momentum in particle collisions and decays without reference to the forces and interactions causing these processes. Electrodynamics obviously defines the dynamics when the forces are electromagnetic in origin.

In his first 1905 relativity paper, Einstein used an electrodynamic approach to obtain the results that gave the formulae for the relativistic energy and momentum of a particle. We shall outline his argument in the next chapter. This chapter will introduce the case of electrically charged particles moving at relativistic speeds in electric and magnetic fields that are constant or only slowly varying in time. We do this using the relativistic formulae we have already met and some simple electricity and magnetism. We shall describe one of the earliest experiments that sought to verify special relativity by measuring the momentum of fast electrons of known speed. A brief description of the steps leading to accelerators capable of reaching relativistic energies will follow. The operation of these devices depends on the application of this simple relativistic electrodynamics.

**4.2 Making contact with electromagnetism

At this point we must establish contact with electromagnetism. Specifically we must examine the forces acting on an electrically charged particle moving under the influence of electric and magnetic fields alone. To do that we begin with elementary electrostatics. The force F_E on a charge q in an electric field E is $F_E = qE$, with F_E in the direction of E. This is correctly expressed in the vector equation

$$\mathbf{F}_E = q\mathbf{E}. \tag{4.1}$$

In elementary magnetism the small element of force dF_M acting on a small element of length dl of wire carrying an electric current I in magnetic induction B is $I B\, dl \sin\theta$ where θ is the angle between the direction of B and the direction of the small length dl. The force is in a direction perpendicular to both B and dl. If $\theta = \pi/2$, that direction is given by Fleming's left hand rule. Expressed in vector notation (Table 3.1) these properties are encapsulated in the vector equation

$$d\mathbf{F}_M = I\, d\mathbf{l} \times \mathbf{B}. \tag{4.2}$$

The direction of $I\, d\mathbf{l}$ is made unique by the direction of flow of the current. Newcomers to vectors should convince themselves that this vector product does give the direction of $d\mathbf{F}_M$ in agreement with Fleming's rule. Since the current is due to the flow of charges (all of one sign) in the wire there must be a force acting on each moving charge in the wire. Suppose there are N such charges q per unit length of wire and each has speed v then we have $I\, dl = Nqv\, dl$. Then the force on each charge is $qvB \sin\theta$. In vector notation this becomes $q\mathbf{v} \times \mathbf{B}$ (Table 3.1). This suggests that we can now put together from equations (4.1) and (4.2) an expression for the force acting on a charge moving under the influence of an electric field and a magnetic induction:

$$\mathbf{F} = q(\mathbf{E} + \mathbf{v} \times \mathbf{B}). \tag{4.3}$$

We have not proved equation (4.3), merely given an argument based upon the existence of magnetic forces on wires carrying electric current. Before Einstein, equation (4.3) was an axiom introduced into electromagnetism by Lorentz.

The units in these equations are coulombs for charge, amperes for current, volts per metre for electric field, teslas for magnetic induction and newtons for force.

We can now draw some simple conclusions concerning the effect of **B**:

1   Suppose a particle with charge q is moving with speed v in a direction momentarily perpendicular to a magnetic induction **B**. The force $q\mathbf{v} \times \mathbf{B}$ acts perpendicularly to **v**. It changes the direction of **v** while remaining perpendicular to **v**. Therefore this force provides a centripetal force that will cause the particle to move in a circle. Non-relativistically that force must be Mv^2/ρ where ρ is the radius of the circle and we have

$$\frac{Mv^2}{\rho} = qvB.$$

Therefore

$$Mv = Bq\rho. \tag{4.4}$$

2   Equation (4.4) is correct in pre-Einstein electrodynamics. But in his first 1905 paper Einstein proved that, for a charged particle moving with any speed less than c, the correct equation is

$$\frac{Mv}{\sqrt{1 - v^2/c^2}} = Bq\rho. \tag{4.5}$$

Einstein did not refer to momentum in presenting this result. However we see from equations (2.1) and (2.3) that the change from (4.4) to (4.5) is the same as replacing $Mv = P$ by $P = Mv/\sqrt{1 - v^2/c^2}$. This last is the relativistic expression for the momentum of a particle. Thus equation (4.5) may be written

$$P = Bq\rho. \tag{4.6}$$

This equation is correct at all attainable speeds for charged particles using the relativistic expression for the momentum.

3   Since the magnetic force is always perpendicular to the velocity of the particle, that force does no work and does not change the particle's kinetic energy. In contrast to that, the electric force is not necessarily perpendicular to the velocity and therefore it can change the kinetic energy.

Figure 4.1 shows some examples of the movement of an electron under some simple arrangements of **E** and **B**. Note that the electron is negatively charged so that q is a negative quantity in the application of equation (4.3).

Equation (4.5) is a particular example of a magnetic induction changing the direction of the momentum of a particle. We can combine equation (4.3) with

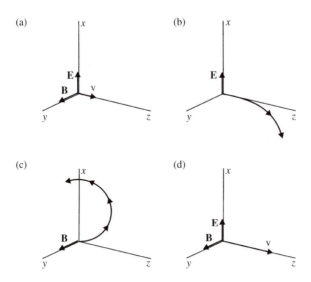

Figure 4.1 Diagrams showing examples of how an electron moves in static and uniform electric and magnetic fields. (a) The electron starts moving from the origin at speed v along the z axis in crossed **E** and **B**. (The non-zero components are E_x and B_y.) (b) How it moves in **E** alone. The motion is in the xz plane. (c) How it moves **B** alone. The motion is in the xz plane. How it moves **B** alone. The motion is in circular in the the xy plane. (d) How it moves if $E/B = v$. In this case the electron suffers no deflection.

Newton's First Law to give

$$\frac{d\mathbf{P}}{dt} = \mathbf{F} = q(\mathbf{E} + \mathbf{v} \times \mathbf{B}).$$

This becomes, using the vector version of equation (2.3)

$$M\frac{d}{dt}\left[\frac{\mathbf{v}}{\sqrt{1 - v^2/c^2}}\right] = q(\mathbf{E} + \mathbf{v} \times \mathbf{B}). \tag{4.7}$$

This is the equation of motion for a charged particle in constant or slowly changing electric and magnetic fields. As we shall find in Chapter 5, equations (4.7) and (4.3) follow from Einstein's results based on his postulates of special relativity.

**4.3   Relativistic mass

There is another way of viewing the momentum–energy relations for a particle. The mass of a particle is considered to vary as

$$M(v) = \frac{M(0)}{\sqrt{1 - v^2/c^2}}, \tag{4.8}$$

where $M(0)$ is the mass at rest. $M(v)$ is called the *relativistic mass* and it varies with the speed. Equations (2.3) and (2.5) then take a simple form:

$$\mathbf{P} = M(v)\mathbf{v}$$

and

$$E = M(v)c^2.$$

Some authors use the relativistic mass but we shall do so only in the next two sections. The reasons for this policy are:

1   The role of this idea in presentations of special relativity is much diminished compared with that in the early days of the subject.
2   The idea places too much emphasis on the use of the velocity or of the speed as an independent variable to define the motion of a particle. This we wish to avoid. We shall use momentum or energy as the situation demands, using equation (2.6) to connect them.
3   Another reason for reducing the importance of the velocity of a particle concerns the equations that transform the components of a velocity between reference frames (Section 7.15). They are not as simple as, or as memorable as, the equations that transform the time–space coordinates of an event (Section 7.2) and the energy–momentum of a particle (Section 6.2).

4.4 Historical interlude

First we recount a situation that existed in the few years before Einstein (1905) published his first relativity papers. While investigating the nature of cathode rays, J. J. Thomson discovered the electron in 1897. Consequently, efforts were made to understand the structure of this particle. A charged body moving through the classical ether (Chapter 10) was expected to have a mass that increases with speed. This variation depends on the nature of the internal charge distribution and on the shape and rigidity of the charge bearing structure. Proposals involved an intrinsic 'material' mass and a variable mass due to the electromagnetic effects. The principal names associated with these efforts are Hendrik Lorentz and Max Abraham. Each had found a formula for the mass increase. Lorentz assumed that an electron was spherical when at rest but when moving suffered a contraction of its diameter along the direction of its motion. The contraction was taken to be the same as was thought to explain the negative result of the Michelson–Morley experiment (Section 10.3). With increasing speed through the ether this contraction increased, and with it the mass. Abraham assumed an electron that was a rigid sphere of charge. The relativistic formula for the momentum of a particle was not given by Einstein in his 1905 relativity papers but, as we shall find, was derived from his results. It can be interpreted as indicating an increase in mass with speed (equation (4.8), Section 4.3). In that context his formula for the mass increase was the same as Lorentz's. Experimental investigations were undertaken to measure how the electron mass varied with its speed, in order to decide which formula was correct and in the hope of shedding some light on the electron's structure.

Consider equation (4.4) for the radius ρ of the circular path of a particle, mass m and charge Q moving perpendicularly to a magnetic induction B:

$$mv = Bq\rho.$$

Applied to the electron we have $q = -e$. The ratio of charge to mass for the electron has been a very significant quantity in the development of atomic physics and has always been written e/m. In this section we represent the electron mass at rest by $m(0)$. Since we are going along with the idea that the mass is varying, we make that explicit and use $m(v)$ to represent the mass at speed v. Then, dropping the negative sign and taking magnitudes, we have

$$m(v)v = Be\rho. \tag{4.9}$$

If Lorentz and Einstein were correct then it was expected that (equation (4.7))

$$m(v) = m(0)\frac{1}{\sqrt{1 - v^2/c^2}}. \tag{4.10}$$

Abraham's formula was

$$m(v) = m(0)\frac{3}{8\beta^3}\left[(1 + \beta^2)\ln\left(\frac{1 + \beta}{1 - \beta}\right) - 2\beta\right], \quad \beta = v/c.$$

The experiments aimed to measure the charge-to-mass ratio $e/m(v)$ as a function of v. The method involved measuring the deflection of a beam of electrons (in a vacuum) by electric and magnetic fields. To extract results it was assumed

1   that the force on the electron was given by the formula proposed by Lorentz, equation (4.3), and
2   that the charge on the electron did not vary with speed.

Thomson had made an order-of-magnitude measurement of e/m for the electron in his work on cathode rays and he did not have sufficient precision to detect an increase of mass at the electron velocities available to him. By 1901 it was known that the radiations from radioactive materials that Rutherford called β-rays were electrons, some having speeds up to $0.9c$. Walter Kaufmann acted on a suggestion that β-rays provided superior candidates for measuring an increase in mass with speed than did cathode rays which were only available at speeds up to $0.3c$. He made a series of measurements of charge-to-mass ratio (and hence of $m(v)$) for β-rays but his precision was worse than he thought it to be. In 1906 he declared the results of his final experiments to agree with the formula for $m(v)$ due to Abraham and not with that due to Einstein and Lorentz. Adolf Bestelmeyer made measurements of the charge-to-mass ratio for electrons ejected by X-rays from a platinum sheet (1906). His method was good but he was limited to electron velocities below $0.3c$ and could not provide verification of one formula over another. Bucherer used the same method as Bestelmeyer but had a strong radium source of β-rays (1908). He determined the charge-to-mass for the electron from $v = 0.3173c$ to $v = 0.6870c$. We describe his method in Section 4.5, and in Figure 4.2.

Einstein's formula for the mass variation with speed is derived from Newton's laws, Maxwell's electromagnetism and the principles of relativity alone. It is independent of the particle's structure or, as Einstein showed, of its electric charge. The fact that with time Einstein was found to be right put an end to any hope of providing information about the structure of the electron by measuring $m(v)$. However, the experimental methods for determining $m(v)$ were a part of the progress towards validating special relativity.

*4.5   Bucherer

Bucherer's electrons came from a strong radioactive source. Such electrons are not mono-energetic and he selected a narrow band of speeds, improving a technique first employed by Thomson. Look at Figure 4.1. In Cartesian coordinates we have arranged the direction of a **B** to be parallel to the y-axis, and an **E** parallel to the x-axis. An electron, charge $-e$ (our convention is that e is the magnitude of the electron charge), moving along the z-axis with speed v, experiences a force **F**

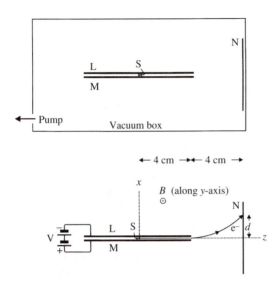

Figure 4.2 This is a sketch of Bucherer's apparatus for the determination of the relativistic mass of electrons as a function of their speed.

- L and M are the plates of a capacitor with separation 0.25 mm. An applied voltage V gives an electric field between the plates.
- S is a radioactive source of β-rays (= electrons). These electrons are not mono-energetic but have kinetic energies up to a maximum characteristic of the source (radium, in Bucherer's case).
- N is a photographic film. Such film is sensitive to fast electrons and so provides a record of the integrated intensity and position of any electron beam that strikes it.
- The whole is immersed in a uniform magnetic field B perpendicular to the plane of the paper and is contained in a vacuum.
- The e^- labels the path of an electron, with speed $v = E/B$, which is travelling along the z-axis. After leaving the capacitor space it is deflected by the magnetic field. The geometry defines a narrow range of speeds and directions. The beam of electrons so defined will mark the photographic film after sufficient exposure.
- A second exposure with both E and B reversed selected the same speed but provided an opposite deflection at N. This allowed a determination of the deflection without a knowledge of the position in the x direction of the photographic plate.
- That deflection gave the value of ρ (equation (4.9)), the radius of curvature of the path beyond the capacitor for the speed selected.
- Apart from the magnetic field B the essential parts of the apparatus were cylindrically symmetric about the axis x. The film N extended $\pm 90°$ from the z-axis. This allowed observation of the deflection for a range of speeds in a single exposure. However, Bucherer used only the deflection for electrons at $0°$ to the z-axis in the capacitor and made a series of exposures with different E and B to select the electron speeds (see Problem 4.1).

(equation (4.3)) that has components given by

$$F_x = -e(E - vB) = -eE + evB,$$
$$F_y = 0, \quad F_z = 0.$$

Thus $\mathbf{F} = 0$ if $v = E/B$. Then electrons from a radioactive source filtered through suitably arranged collimators and crossed static electric and magnetic fields can yield a beam having speeds within a small, known range.

Bucherer's velocity selected electrons then traversed a region where there was no electric field but the same magnetic field. Here they followed a part of a circle of radius ρ given by equation

$$m(v)v = Be\rho.$$

Then, if the Einstein/Lorentz formula was correct (equation (4.10))

$$\frac{e}{m(0)} = \frac{v}{\rho B \sqrt{1 - v^2/c^2}}, \tag{4.11}$$

where $v/c = E/B$. We have expressed the result in the e/m form as this is what Bucherer did. A sketch of his apparatus is shown in Figure 4.2 and in Table 4.1 we show his results. He measured $e/m(v)$ at five speeds (Cols 1 and 2). Using equation (4.11) he calculated $e/m(0)$ (Col. 3). The last column (4) shows the results of a similar calculation, but based upon Abraham's formula for the mass variation. Figure 4.3 shows these results graphically.

Try Problem 4.1 to obtain an idea of the magnitudes of the quantities involved in Bucherer's work.

Table 4.1 Bucherer's measurements for the charge-to-mass ratio of the electron (Col. 2) at different speeds (Col. 1). Column 3 shows the values of $e/m(0)$ expected from each measurement for Einstein's formula for $m(v)$. Column 4 shows the same for Abraham's formula. The numbers in Column 3 agree within 1%. The numbers in Column 4 show a nearly systematic fall of about 5%. The modern value for $e/m(0)$ is $1.7588 \times 10^{11}\,\mathrm{C\,kg^{-1}}$

v/c	$e/m(v)$ measured	$e/m(0)$ Einstein $(10^{11}\,\mathrm{C\,kg^{-1}})$	$e/m(0)$ Abraham
0.3173	1.661	1.752	1.726
0.3787	1.630	1.761	1.733
0.4281	1.590	1.760	1.723
0.5154	1.511	1.763	1.706
0.6870	1.283	1.767	1.642

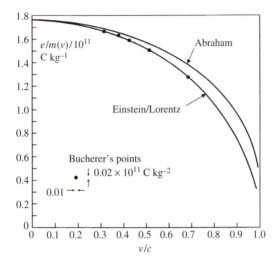

Figure 4.3 Bucherer's results for $e/m(v)$ for the electron plotted (●) against v/c and compared with Einstein's and with Abraham's formulae, both calculated using the modern values of e and $m(0)$. The diameter of the plotted points (●) for Bucherer's results correspond to 0.02 and 0.01 in the values of $e/m(v)$ and of v/c respectively.

Bucherer gave no errors beyond those implied by giving his results to four significant figures. Each of his five results for $e/m(v)$ when scaled back to rest using Einstein's formula, are within 0.5% of the modern value, which he could not have known. Scaling back using Abraham's formula gives results below the modern value by 2–7% (Table 4.1). The results confirmed Einstein's formula.

That was Bucherer's conclusion. However, as late as 1957 there were criticisms of these and of later measurements, on the grounds that the precision claimed or implied was not adequate to decide between Abraham and Einstein. To physicists of the late twentieth century it is strange that there was a contest between Einstein and Abraham. The latter's formula for the increase in mass was derived assuming a non-point-like electron moving through an ether. We may not have a full understanding of the electron but it is certainly not an extended object of the kind that Abraham or Lorentz may have envisaged, and the ether has long been abandoned.

It is incorrect to judge as redundant a whole series of measurements of $e/m(v)$ following Bucherer, on the criterion of discriminating between Einstein and Abraham. Increasing precision helped to persuade doubters of special relativity. No experiment disagreed with Einstein and it seems he never had any doubts.

*4.6 Accelerators

These are devices built to provide well-controlled beams of energetic charged particles. The first accelerator used to investigate artificially induced nuclear reactions (1932) was built by J. D. Cockcroft and E. T. S. Walton. We have already discussed one of the early results found by these authors (Section 2.7). Protons were accelerated by a high DC voltage generated in a rectifier of special design. Such DC machines were limited by the difficulties of generating and handling very high voltages. Developments that followed Cockcroft and Walton's 750 kV allowed the acceleration of singly charged particles up to about 20 MeV in DC machines.

One route of development beyond the few MeV of DC machines was with circular machines. The charged particles are constrained to move in near-circular orbits by magnetic fields. Acceleration is provided by repeated small impulses from an applied alternating voltage. The frequency is set by the time taken for particle to traverse the orbit and the phase is such that the impulses are uni-directional. Now momentum and speed are linked in the operation of such devices which thereby provide a test of special relativity, particularly as the speeds attained approached the speed of light. To examine this matter we first we must look at the *cyclotron frequency*.

*4.7 The cyclotron frequency

Look at Figure 4.1 and, in particular, the case of an electron in a magnetic field alone. Assume the magnetic field is uniform and parallel to the y-axis. In your mind's eye change the electron to a proton (charge $+e$) that starts from the origin with momentum $\mathbf{P}$, velocity $\mathbf{v}$, both along the z-axis. Subsequently it will traverse a circle, radius r, lying in the xz plane with centre at $x = -r$, $y = 0$, $z = 0$. Then, at non-relativistic speeds we can use equation (4.4):

$$Ber = Mv$$

and the period of revolution of the circle T is given by

$$T = 2\pi r/v = 2\pi P/(Bev).$$

Then the **cyclotron frequency** $f_c \equiv 1/T$ is given by

$$f_c = Bev/(2\pi P). \tag{4.12}$$

For non-relativistic speeds $v \ll c$, $P = Mv$, and

$$f_c(v \ll c) = Be/(2\pi M). \tag{4.13}$$

Note that this frequency is independent of P and of r.
For relativistic speeds we expect $P = Mv/\sqrt{1 - v^2/c^2}$. Then equations (4.12) and (4.13) give

$$f_c(v) = f_c(v \ll c)\sqrt{1 - v^2/c^2}. \tag{4.14}$$

Thus, in a uniform magnetic field the period of revolution $(1/f_c)$ increases as the speed increases.

*4.8   The cyclotron

E. O. Lawrence and M. S. Livingston at the University of California at Berkeley, were the first to build a magnetic resonance accelerator, or **cyclotron** as it came to be labelled (1931). Figure 4.4 shows a schematic of the device. Ions circulate inside a vacuum box in a plane perpendicular to a nearly uniform, constant magnetic field. The ion orbits lie inside a divided, shallow cylindrical box made of copper. The division is across a diameter. The two halves constitute two electrodes (called dees) to which an alternating voltage at the cyclotron frequency is applied. (This voltage is normally called the Radio Frequency (RF) voltage because the frequency was typical of short wave radio. The techniques for its generation and handling at high power were well established.) The ions are produced in a central source and come under the influence of this RF voltage which appears across the gap between the

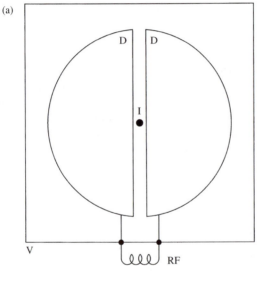

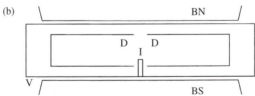

Figure 4.4 The essential parts of a cyclotron. Part (a) shows a plan view in a plane perpendicular to the magnetic field and part (b) an elevation through the axis and perpendicular to the gap between the dees (D). (1) BN and BS are pole tips of the magnetic field; (2) I is the ion source; (3) RF is the source of the RF voltage; (4) V is the vacuum box kept at low pressure by a pump.

dees. Some ions are caught in the acceleration process. These follow near circular orbits such that they always cross the gap between the two dees at a time when the electric field at the gap gives the particles a push. Thus they follow a slowly expanding spiral path. At maximum radius, the accelerated ions may be extracted or allowed to strike an internal target for the purposes of research. Note that the accelerated beam will consist of a continuous stream of short bursts of nearly mono-energetic ions separated by a time $1/f_c$.

Lawrence's first cyclotron was small (diameter of 10 cm) but by 1941 he had built several and reached a diameter of 1.5 m with a final accelerated proton energy of 8 MeV. In each the frequency was held constant and close to the expected cyclotron frequency. However, the detailed design had to take account of a small increase of the period of revolution due to the relativistic effects and to a built-in, slow decrease in the magnetic field with increasing radius. This decrease was deliberately designed to provide focusing of the beam during acceleration. The success of these machines could not do more than confirm the correctness of equations (4.3) and (4.4) for non-relativistic speeds.

*4.9 The limitations of the cyclotron

Consider a small cyclotron with $B = 0.5\,\mathrm{T}$ ($=$ tesla), maximum $r = 0.5\,\mathrm{m}$, accelerating protons. Then $f_c = 7.623\,\mathrm{MHz}$. At maximum radius $v/c = 0.07963$ and the kinetic energy is 2.99 MeV. The relativistic effect reduces the frequency of circulation at maximum radius by 0.32% to 7.598 MHz. The cyclotron at fixed frequency can deal with this change. The RF voltage has to be sufficient to provide acceleration rapid enough to prevent the ions falling back in phase with respect to the RF so as to stop acceleration.

Now try $B = 1.0\,\mathrm{T}$, $r_{max} = 1.5\,\mathrm{m}$, accelerating protons. The cyclotron frequency $f_c(v \ll c) = 15.25\,\mathrm{MHz}$. At maximum radius the momentum of the protons is 450 MeV/c, the kinetic energy is 102 MeV and $v/c = 0.432$. At this speed the cyclotron frequency is

$$f_c(v/c = 0.432) = f_c(v \ll c)\sqrt{1 - v^2/c^2}$$
$$= 15.25 \times 0.902 = 13.75\,\mathrm{MHz}.$$

(Problem 4.2) In addition there would be the effect of the decrease in magnetic field at increasing radius. The change of circulation frequency from 15.25 to less than 13.75 MHz during acceleration would make acceleration impossible at constant applied RF without excessively high RF voltages.

Designer Lawrence was warned by theoretical physicists that special relativity would prevent the operation of cyclotrons at ion speeds not small compared with the speed of light. He saw that as a challenge to find ways of providing very high

RF voltages on the cyclotron dees. However that was proved to be unnecessary by the discovery of phase stability, described in the next section.

4.10   The synchrocyclotron

This device depends on the principle of **phase stability**, proposed independently by V. I. Veksler (USSR) and E. M. McMillan (Berkeley) in 1945. Consider a bunch of ions circulating in a cyclotron with a speed that is a significant fraction of the speed of light. Suppose that, on average, members of the bunch are neither gaining nor losing energy. A possible motion would be one in which a single ion always crossed the gap between the dees at an instant when the RF field was passing through zero. This ion would remain at the same radius. Suppose the phase of this zero with respect to the bunch crossing is such that the field is changing from accelerating to decelerating. Members of the bunch of a lower speed than average have a higher cyclotron frequency; they will arrive early for their next gap crossing and will receive a small accelerating push. Conversely faster members would receive a decelerating impulse. Thus members are prevented from straying significantly from the average energy. This is phase stability. Now slowly reduce the applied RF. Stability means the bunch must keep its circulation frequency matching the decreasing RF. The bunch progresses to higher momentum and greater radius. Thus a bunch can be escorted from small radius and low energy to the maximum radius and maximum energy without large RF voltages. The average energy gain per gap crossing is now small and there are many more crossings in the acceleration process than in a conventional cyclotron.

Thus a **synchrocyclotron** has a magnetic field that is constant (in time), decreasing slightly with increasing radius out to the maximum radius for acceleration. The RF accelerating system operates with modest voltages. A bunch of ions that start from the centre at the cyclotron frequency, is guided to maximum radius as the applied frequency decreases, and then extracted, or allowed to strike an internal target. This acceleration cycle is repeated at about 200 Hz.

Many proton synchrocyclotrons were built, worked well, and had very useful research lives. To name a few: Lawrence's 350 MeV, later 720 MeV, at the University of California at Berkeley, 400 MeV at the University of Liverpool, 450 MeV at the University of Chicago, 600 MeV at CERN, in Switzerland, and 680 MeV at The Joint Institute for Nuclear Research, at Dubna in Russia. Try Problem 4.3 to obtain an idea of the parameters of a large synchrocyclotron.

Before the operation of the first synchrocyclotron, studies of cosmic radiation at high altitudes had led to the discovery of new particles. One of these, initially called the π-mesotron, now called the pion (Section 3.2), was found to be produced readily in collisions of very energetic cosmic rays (mostly protons) with atomic nuclei in the upper atmosphere. The first artificially produced pions were observed with the Berkeley synchrocyclotron in 1948. Thereafter synchrocyclotrons served in extensive investigations into the properties of these particles.

At this stage we should note that the frequency range over which the RF of the synchrocyclotrons had to be modulated agreed with the formula of special relativity (equation (4.14)). Good marks to Einstein for giving us the means to predict correctly the momentum of protons out to $v/c \approx 0.81$. However, phase stability means that the orbit of accelerated particles was self-adjusting to the RF. At each frequency the orbit radius could not be precisely known because of beam dynamics. This dynamics includes vertical (parallel to the magnetic field) and horizontal oscillations and oscillations in phase with respect to the RF. A precision test of special relativity becomes difficult and none has been published.

The cost of providing a nearly uniform magnetic field out to diameter of 6 m (Dubna) prohibited further increases in energy using the synchrocyclotron.

4.11   The synchrotron

The principle of phase stability is now applied to particle orbits of constant average radius. In the **synchrotron** a magnetic field is provided only locally at this radius but has to be increased during the acceleration cycle. The accelerating RF voltage now has to be frequency modulated as the speed and the momentum of the particles increase. The tracking of this frequency with the increasing magnetic field is governed by the need to keep the orbit radius constant.

Several proton synchrotrons have been built and operated in the USA, in Europe, and in the USSR. The earliest included a 1 GeV machine at the University of Birmingham in the UK, the 3 GeV Cosmotron at the Bookhaven National Laboratory and the 6 GeV Bevatron at Berkeley, both in the USA. Developments in the techniques for controlling the dynamics of the beam during acceleration permitted savings in magnet costs and further increases in energy. The highest energy synchrotron is the Tevatron at the Fermilab in the USA which is capable of accelerating protons to an energy close to 1 teraelectronvolt (TeV) (= 1000 GeV). The Cosmotron and Bevatron allowed the first artificial production and controlled observation of the lightest strange particles. During the five decades that followed 1950, the increasing energies available for research led to an explosion in our knowledge and understanding of elementary particle physics. However, the immediate question concerns the precision and range of validity added to the verification of Einstein's momentum–velocity relation (equation (2.3)) by the successful operation of very high energy synchrotrons.

The protons in the Tevatron traverse an orbit that passes through 774 magnetic dipoles and 216 magnetic quadrupoles. The effective magnetic field along the average orbit at the end of acceleration will give the final momentum (as in equation (4.6)). The frequency of circulation can be measured with very high accuracy and that allows a determination of the speed. At 1000 GeV that speed is $c(1 - 4.4 \times 10^{-7})$. However, the reciprocal of $\sqrt{1 - v^2/c^2}$ prevents an accurate measurement of the speed becoming a precise determination of the right-hand side of equation (2.3). A good test of the relativistic formula for the momentum of a particle as a function of its speed becomes very difficult, although it is clear that the final speed does not exceed c.

4.12   Electron accelerators

There are many types and all operate correctly assuming electrons with a kinetic energy of a few MeV and above move imperceptibly less than the speed of light. The highest energy reached for electrons (and positrons) is just over 100 GeV at the CERN electron–positron collider (LEP, Section 4.13). At this energy the speed is expected to be $c(1 - 1.31 \times 10^{-11})$. There is no evidence that c is exceeded. So electron accelerators confirm the speed limitation on particles with rest mass.

There is one serious complication in the operation of circular accelerators for electrons (or positrons). In the magnetic field that keeps the electrons in a closed orbit there is an acceleration experienced by these particles towards the centre of curvature. This acceleration causes the electrons to radiate photons close to their direction of motion. This process is called the emission of *synchrotron radiation*. It is an energy loss mechanism that becomes greater with increasing electron energy at fixed orbit radius. The RF accelerating power must be increased above that required in the absence of the effect, to compensate for the loss. The Large Electron–Positron (LEP) collider at CERN (Section 4.13) has an energy loss per turn per particle of about 2 GeV at 100 GeV. At the level of presently foreseeable technology this means that LEP will have been the highest energy circular electron and positron accelerator.

The energy radiated by each particle per turn at a given radius varies as $(E/M)^4$. Therefore synchrotron radiation is of no importance in existing circular proton accelerators; however, the Large Hadron Collider (LHC) under construction at CERN (Section 4.13) will have a significant energy loss per turn of about 5 keV.

4.13   Centre-of-mass energy at colliders

Look again at equation (3.24); it gives us W, the centre-of-mass energy, for collisions of extreme relativistic particles of energy E_a striking a fixed target containing particles of mass M_b. We note that W only increases as does the square root of E_a. One measure of the potential of an accelerator is W calculated for a fixed target of protons. The financial cost of accelerators may increase as E_a or even faster. Thus increasing expenditure gives poor returns as far as increases in centre-of-mass energy are concerned. This, and other reasons that will become evident, led to the development of colliders.

A collider is a device in which two beams of particles intersect at an angle such that their momenta are oppositely or nearly oppositely directed. At the intersection region there are collisions between a particle from one beam with a particle from the other beam. The technical achievement of this situation is due to the skill and devotion of many accelerator designers. However, that is another story. We are concerned with the kinematics of these collisions.

Equation (3.18) gives us W for collider collisions. Let a particle from beam 1 have energy and momentum E_1, $\mathbf{P}_1$ and a particle from beam 2 have energy and momentum E_2, $\mathbf{P}_2$. Then

$$W = \sqrt{(E_1 + E_2)^2 - (\mathbf{P}_1 c + \mathbf{P}_2 c)^2}. \tag{4.15}$$

For fixed E_1 and E_2 the greatest W will occur if the two momenta are in opposite directions so that $(\mathbf{P}_1 c + \mathbf{P}_2 c)^2 = (P_1 c - P_2 c)^2$. (Try Problem 4.4.)

From small beginnings by pioneers, colliders have grown in size and in the impact they have had on the physics of elementary particles. Some examples of the most energetic follow:

1   The LEP collider at CERN, Geneva, Switzerland: $E_1 = E_2 = 104\,\text{GeV}$, $\mathbf{P}_1 = -\mathbf{P}_2$, $W = 208\,\text{GeV}$ (year 2000 data). Operation ceased in 2000 to make way for the construction of the LHC in the same tunnel as occupied by LEP.
2   The Tevatron at the Fermilab, USA. A proton–antiproton collider with $E_1 = E_2 = 1000\,\text{GeV}$, $\mathbf{P}_1 = -\mathbf{P}_2$, $W = 2000\,\text{GeV}$ (2000 data).
3   The HERA collider at the DESY Laboratory, Hamburg, Germany. This device collides protons with electrons or positrons. $E_p = 820\,\text{GeV}$, $E_e = 30\,\text{GeV}$, $|\mathbf{P}_1 + \mathbf{P}_2| = (820 - 30)\,\text{GeV}/c$, $W = 314\,\text{GeV}$ (1999 data).
4   At CERN the next collider is under construction. Called the LHC it is due to start operation in the year 2005. It will be a 7 TeV on 7 TeV proton–proton collider (recall 1 TeV = 1000 GeV). Therefore $W = 14\,\text{TeV}$, seven times the Tevatron collider's W.

[A full list may be found in the *Review of Particle Physics*.[1]]

There is an interesting quantity that may be calculated for each collider. Find the energy of an incident particle that, in a collision with a stationary particle, reaches the same centre-of-mass energy as in the collider. For the fixed target imitator of the Tevatron we have (equation (3.24))

$$W^2 = 2\,E M_p c^2,$$

where $M_p c^2 = 0.938\,\text{GeV}$ (target proton), $W = 2000\,\text{GeV}$, E is the energy of incident anti-proton required to reach this W. The result is

$$E = 2,130,000\,\text{GeV}.$$

This is over two thousand times the particle energy in one beam of the Tevatron collider which in itself is the most energetic proton accelerator yet built. Such a huge step up in energy is impossible with existing techniques and resources. These numbers highlight the importance of colliders. Now do the same calculations for LEP and HERA: Problem 4.5.

*4.14   Summary

The design and construction of particle accelerators and colliders to ensure beam stability and focusing during acceleration and storage is a critical matter. It is done

---

1 Particle Data Group, *European Physical Journal* **C5**, 2000, 160–2.

assuming the equation (4.7) is correct. Nothing has happened to suggest that this is the wrong thing to do in the design either of proton accelerators for energies up to 1000 GeV or of electron accelerators up to 100 GeV. In addition, no accelerator has succeeded in accelerating beyond the speed of light.

Although accelerators have not provided precision tests of the momentum–speed relation at all accessible speeds, their success has helped validate the electrodynamic formulae that follow from Einstein's postulates of special relativity. It is the physics research that exploited the potential of accelerators that has amply verified the relations.

Why do we strive for higher and higher centre-of-mass energies? Since the first artificially produced pions were detected in 1948, the increase in the available collision energies has lead to many discoveries and to a deepening of our understanding of the behaviour of matter. There is every reason to believe that this will continue. The LHC at CERN will possibly allow the discovery of a predicted particle, the Higgs boson, if it exists. The Higgs' mass is expected to be about $100 \, \text{GeV}/c^2$, which is well within the planned energy capabilities of the LHC.

Problems

4.1   The largest speed of electrons for which Bucherer measured e/m was $0.6870c$. What was their kinetic energy (in MeV) and their momentum (in MeV/c)? Assuming he set the magnetic field B in his apparatus (Figure 4.2) to be 10^{-2} T, find

    a the electric field between the plates of the capacitor,
    b the voltage difference between the plates, and
    c the deflection (d in Figure 4.2) of these electrons at the photographic plate.

4.2   Check the values given in Section 4.9 for energies, momenta, speeds, and cyclotron frequencies in the examples of two cyclotrons of different radii.

4.3   Consider a synchrocyclotron designed to accelerate protons to a kinetic energy of 680 MeV and having a uniform magnetic field of 1.5 T.

    a Find the radius of the final orbit.
       Calculate the cyclotron frequency for the protons
    b at the beginning of the acceleration cycle, and
    c at the end of the acceleration cycle.

4.4   The world's largest accelerators give or have given:

    a CERN Super Proton Synchrotron (Switzerland): 450 GeV protons.
    b Fermilab Tevatron (USA): 1000 GeV protons.
    c The LEP collider (CERN, Geneva). 104 GeV electrons on 104 GeV positrons.
    d Electron–proton collider HERA at Hamburg (Germany): 30 GeV electrons on 820 GeV protons.

Calculate the largest rest mass for particle X (in GeV/c^2) that could be produced in the following processes:

1 For (a) and (b) in fixed target operation:

$$p + p \rightarrow p + p + X.$$

2 For (c)

$$e^+ + e^- \rightarrow W^+ + W^- + X,$$

where the W^+ and W^- are known particles, both of mass $80.4 \, GeV/c^2$.

3 For (d)

$$e^- + p \rightarrow e^- + p + X.$$

4.5 See Section 4.13.

a The LEP collider operated at CERN in Switzerland. The energy parameters are given in Problem 4.4(c). Find the laboratory energy positrons would have to have so that striking electrons at rest, the colliding system had the same centre-of-mass energy.

b The HERA collider parameters are given in Problem 4.4(d). (i) Find the energy of electrons necessary to reach the same centre-of-mass energy in collisions with stationary protons. (ii) Do the same for protons striking stationary electrons.

5 The foundations of special relativity

**5.1 Introduction

Chapters 2–4 should have given the reader a basis for accepting that special relativity is easy to use. Now we have to return to its origins to describe how it was discovered and developed. We do this by mapping out the journey, leaving a detailed examination of the important steps to later chapters.

All readers will be familiar with falling rain. To a stationary observer the rain might be falling vertically downwards. To an observer moving on a bicycle that rain will appear to be falling at an angle. However, if a steady wind is blowing, the stationary observer might see the rain falling at an angle and the cyclist, if he is moving with the wind, will see the rain falling vertically. In these particular circumstances there is no difficulty in deciding who is stationary and who is moving with respect to the ground. Remove everything but the two observers and replace the rain by a beam of light from a very distant star. To one observer the second could be approaching at constant speed. In that case, to the second observer, the first is also approaching with the same speed. The star may be so distant, or absent, that it is impossible to determine which observer is stationary and which is moving. The apparent direction of the star with respect to the line joining the observers may be different for the two observers but that does not help. In fact, all that may be determined is the relative speed between the two observers.

Relativity is about the consequences for the laws of physics that follow from the principle that it is impossible to detect a state of uniform motion of an isolated system. It is only the relative motion of two systems that is observable. This is the **Principle of Relativity**. It is the basis of all that follows. Galileo Galilei (1564–1642) was the first to publish a formulation of a principle of relativity.[1]

To find the implications of the principle of relativity we have to investigate the nature of the relation between the result of a measurement of a physical quantity by an observer and the result of a measurement of the same physical quantity by a second observer moving uniformly relative to the first. By the 'same' quantity we mean a quantity that is not necessarily associated with either observer but may

---

1 A relevant quotation, translated from his writings, may be found in Taylor, E. F. and Wheeler, J. A., *Spacetime physics*, 2nd edn, San Francisco: W.H. Freeman & Co, 1992, pp. 53–4.

be observed by both. Two examples are the velocity of a single particle or the coordinates of an event. There are also **proper quantities** that are associated with each observer. Examples are the wavelength of a spectral line from an agreed design of lamp stationary relative to an observer and the mean life of atoms of a radioactive source similarly stationary. From many observations each observer will formulate some laws of physics.

A principle of relativity requires:

a Proper quantities must be the same for both observers, given agreement on the basic units of measurement.

b The laws formulated by the observers must be identical in their content and contain no feature that allows either observer to claim the presence of a self motion which is other than relative to the other observer.

In Section 5.2 we give some definitions and describe some terms to be used. In Section 5.3 we describe the relativity that applies in classical (pre-Einstein) mechanics. In Section 5.4 we meet the difficulties of reconciling that relativity with electromagnetism. Sections 5.5–5.9 introduce and outline the first steps in Einstein's relativity.

**5.2 Some definitions

To make physical observations, in particular of the motion of bodies, it is necessary to have a system of coordinates and of clocks. We have already referred to this as a **reference frame** (Section 3.1). If the motion of a body observed using this reference frame obeys Newton's First Law of Motion then this is an **inertial reference frame**. We shall frequently shorten *inertial reference frame* to **inertial frame**. Any inertial frame has the following properties.

1 Any body, in such a frame and not acted on by any forces external to itself, continues in a state of rest or of uniform motion in a straight line (constant velocity). This is Newton's First Law and, as we have indicated, it is a characteristic of any inertial frame.

2 An inertial frame cannot be accelerating or rotating, and gravitational fields must be absent.

3 Newton's Second and Third Laws also hold in all inertial frames.

(2nd) The time rate of change of momentum (P) of a body is equal to the total force (F) acting upon it. That is

$$\frac{dP}{dt} = F. \tag{5.1}$$

(3rd) To every force there is an equal and opposite reaction.

4   There is no way of defining a unique or preferred frame that can be said to be at rest or in some state of uniform motion.
5   Thus only the **relative** motion of two inertial frames may be observed. Hence the word *relativity* appears in the name of our subject.
6   The relative motion of any pair of inertial frames must be, by definition, uniform.

A body not subject to any force will not appear to move in a straight line if the reference frame in which it is situated is accelerating. To an observer stationary in that reference frame, the body will appear to be moving with increasing speed in a particular direction and will not to be in a state of uniform motion. A body cannot remain at rest away from the axis of rotation of a rotating reference frame unless there is an appropriate centripetal force acting upon it. By definition, no such force can exist on a body at rest in an inertial frame. Thus acceleration and rotation of inertial reference frames are excluded.

Why must gravity be absent? For example, the weightlessness observed in artificial earth satellites might be thought to indicate an inertial frame. But not so. There are tide-raising forces acting on the seas and material of the Earth due to the gravitational attraction between the Earth and Moon and to their rotation about their mutual centre-of-gravity. Similarly, there are tidal forces acting on a body displaced from the centre-of-gravity of the satellite. The centres-of-gravity of two satellites not in the same circular orbit do not have a uniform relative motion. A frame pinned to the surface of the Earth is not an inertial frame although experiments designed to test special relativity may be performed in such frames. However, the effects due to gravitation and to the acceleration suffered in Earth-bound experiments will be negligible in the cases that we shall consider. We can safely interpret such tests as if they were performed in an inertial frame. So, we ignore gravity. (Of course Einstein did not. After special relativity he tackled gravity and created the General Theory of Relativity.)

An **event** is something that happens at some point in space and at some instant of time.

A **clock** is a device that records the passage of time. All our clocks are identical. For example, each could be consistent with the Standard International Unit in that one second of elapsed time is signalled by 9,192,631,770 periods of a particular microwave transition in the atomic isotope caesium-133.

A clock may be accelerating. It is said to be an **ideal clock** if an infinitesimal time interval it records is the same as that recorded on an ordinary clock stationary in an inertial frame instantaneously moving with the ideal clock.

Distance is measured in metres (SI unit). One metre is defined as the distance moved by light in a vacuum during an interval that is a fraction $1/c$ of one second, where $c = 299,792,458$. By this definition, the **speed of light** is therefore exactly c metres per second. We use c to represent the speed of light (and we *always* mean in a vacuum). To three significant figures $c = 3.00 \times 10^8 \, \mathrm{m \, s^{-1}}$. However, we should not use the speed of light to define the unit of distance unless and until we are convinced that the speed of light is the same in all inertial frames.

The **coordinates** of an event in a given inertial frame are x, y, z (we are using Cartesian coordinates), to which we add t, the time of the event. We shall deal later with the definition and meaning of this time. In any other inertial frame the coordinates of this event are not, in general, the same.

An **inertial observer** is the person who makes a record of the coordinates of events that occur in their inertial frame. This observer is at rest and has observing equipment, including clocks, throughout the inertial frame. The equipment is designed to send to the observer the coordinates x, y, z **and** t of any event. The observer will also have the power to record other quantities such as velocities, momenta and energies of particles, and so on. All our observers will be *inertial*.

A **transformation** is the set of equations for changing quantities observed or equations formulated in one inertial frame to their values observed or equations formulated in another inertial frame.

An **invariant** is a quantity that has the same value independent of the inertial frame in which it is measured. For example, electric charge, mass at rest and, in special relativity, the speed of light are all invariant. We shall frequently draw attention to quantities that are invariant.

Physical laws that are the same in all inertial frames are also said to be *invariant*. The mathematical equations describing such laws will have the same form (that is, the same physical meaning irrespective of notation) in all frames and are said to be **covariant** or **form invariant**. Maxwell's equations are covariant in special relativity.

A **particle** is the term we have already used for what is more properly called a *mass point*. We use *particle* because *elementary particle* is the label frequently given to the sub-atomic constituents of matter, such as the electron and the proton. These behave like mass points as far as the kinematics of special relativity is concerned. The mass has to be at a point because the concept of an extended rigid body is not consistent with special relativity. The reason is that a movement on one side of a *rigid* body is transmitted instantly to the other side. This is impossible in special relativity since information about movement at one place in a body can not be conveyed to another part of the body faster than the speed of light.

A **conservation law** asserts that a given quantity is unchanged in certain processes. Total electric charge, total momentum, and total energy are conserved in collisions between particles. These particular cases are true in Newtonian mechanics and in special relativity.

We shall frequently use a particular arrangement of two inertial reference frames that are in a state of uniform relative motion. Each reference frame is a Cartesian coordinate system. The arrangement is shown in Figure 5.1. A right-handed Cartesian coordinate system (coordinates x, y, z) is one frame and is called Σ. A second, Σ' (x', y', z'), is moving with velocity u in Σ. This velocity is in the direction of the z-axis of Σ. There is an event at which the origins (O and O') of these two coordinate systems coincide and the axes x', y', z' lie along the axes x, y, z respectively. Thereafter, the origin O' of Σ' moves along the z-axis of Σ with speed u. And, apart from the increasing separation of the origins, the z'- and z-axes continue to coincide. We call this the **standard arrangement** of two inertial frames. Any other arrangement of the orientation of the axes or of the position of the origin of either reference frame may be transformed to the standard arrangement. This can be done by a rotation and a displacement of one or both of the coordinate systems. Such operations may be mathematically involved but have no effect on the conclusions that may be drawn from the application of relativistic principles.

We shall describe some other important terms on the way through the subject. The keywords will be printed in boldface.

**5.3   Galilean transformations

In this section we describe briefly the transformations between inertial frames that satisfy the assumptions that Newton made about time and space. These are:

1   The time interval between two events is the same for all observers and independent of the inertial frame in which each observer is located. This is the same as saying that there is a universal time.
2   The spatial separation of two events is the same for all observers acting under the same conditions as in 1.

We consider two inertial frames as in the standard arrangement shown in Figure 5.1. The **Galilean transformations** connect the coordinates x, y, z, t in Σ of an event with the coordinates x', y', z', t' in Σ' of the same event. If we agree about a common zero for the universal time we have by definition

$$t = t'. \tag{5.2}$$

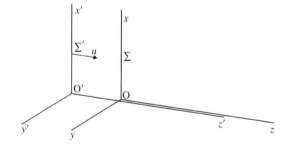

Figure 5.1 The standard arrangement of two inertial reference frames. This sketch shows the two right-handed Cartesian coordinate systems Σ and Σ'. The relative velocity u is shown by an arrow attached to the x'-axis (of Σ'). This means that the origin of Σ' (the point O′) is approaching the origin of Σ (the point O). At the instant that O and O′ coincide, the three axes of one inertial reference frame coincide with the three axes of the other. Thereafter the distance from O to O′ is increasing with time.

If we agree that at $t = t' = 0$ the origins and corresponding axes of Σ and Σ' coincide then, by simple deduction (Figure 5.1),

$$x = x',$$
$$y = y', \tag{5.3}$$
$$z = z' + ut'.$$

Consider now the case of two events, A and B, separated by $\Delta x, \Delta y, \Delta z, \Delta t$ in Σ and $\Delta x', \Delta y', \Delta z', \Delta t'$ in Σ'. Then from equations (5.2)–(5.3) we have

$$\Delta t = \Delta t', \tag{5.4}$$

and

$$\Delta x = \Delta x', \quad \Delta y = \Delta y', \quad \Delta z = \Delta z'.$$

The spatial separation ΔS of the two events in Σ is given by

$$
\begin{aligned}
\Delta S &= \sqrt{(\Delta x)^2 + (\Delta y)^2 + (\Delta z)^2} \\
&= \sqrt{(\Delta x')^2 + (\Delta y')^2 + (\Delta z')^2} \\
&= \Delta S' \text{(the separation in } \Sigma'\text{).}
\end{aligned}
\tag{5.5}
$$

Thus equations (5.4) and (5.5) show that these transformations satisfy Newton's assumptions.

If these spatial coordinates of equations (5.3) are the instantaneous position of a moving particle at a time t we have that the components of its velocity satisfy (remembering u is constant)

$$\frac{dx}{dt} = \frac{dx'}{dt'}, \tag{5.6}$$

$$\frac{dy}{dt} = \frac{dy'}{dt'}, \tag{5.7}$$

$$\frac{dz}{dt} = \frac{dz'}{dt'} + u. \tag{5.8}$$

Thus the z and z' components of the particle's velocity differ by u, as common sense says they should. Light travelling with speed c along the axis z' in Σ' should, by the same reasoning, have speed $c + u$ when observed in Σ. Experimentally that is not the case: the speed of this light is the same in both inertial frames. So we must be prepared for some modifications to Newton's assumptions.

Further differentiation of equations (5.6)–(5.8) shows that any acceleration of the particle is the same in all frames:

$$\frac{d^2x}{dt^2} = \frac{d^2x'}{dt'^2}, \tag{5.9}$$

and so on for all three components, provided u is constant, as it must be for two inertial frames.

We can check that Newton's laws are invariant under Galilean transformations:

1   Since u is constant, equations (5.6)–(5.8) tell us that uniform motion in Σ' is uniform in Σ.

2   Equation (5.9) and two other equations for the two remaining components tell us that the acceleration is the same in Σ' and in Σ. We expect the forces existing between any two bodies to depend solely on the distance between them. This distance and therefore also the forces are the same in Σ' and in Σ. Assuming the invariance of mass, from these properties it follows that Newton's Second Law is invariant.

3   If in some frame (Σ' say) the forces acting between the two bodies are the equal and opposite reaction of Newton's Third Law, then the argument made in the last paragraph shows that these forces observed in Σ are also equal and opposite. Thus Newton's Third Law is invariant.

This invariance of the laws of classical mechanics under Galilean transformations can be expressed in the statement that '*in classical mechanics all inertial frames are equivalent*'. This is sometimes known as the **principle of Galilean relativity**.

5.4   Electromagnetism and Galilean relativity

We have to discuss the relation between electromagnetism and Galilean relativity as it appeared in the last decades of the nineteenth century. To avoid entanglement in a now retired system of units, we do this using SI units.

Coulomb's law for the force between two electric charges in a vacuum is

$$F_{\text{E}} = \frac{q_1 q_2}{4\pi \varepsilon_0 d^2} \quad \text{(newton)}, \tag{5.10}$$

where q_1, q_2 are the charges and d is their separation. The constant ε_0 (*the permittivity of free space* or *the electric constant*) is there because there is no pre-existing knowledge, apart from experimental results, about the force in newtons between charges measured in coulombs at a separation measured in metres.

Consider two infinitely long electrically conducting wires that are straight and arranged to be parallel in a vacuum a distance d apart. Suppose one is carrying a current I_1, the other a current I_2. Then there is a force acting on each wire due to the current in the other. The force acts perpendicularly to the wire and towards or away from the other wire, depending on the relative directions of the currents. The magnitude of this force, expressed as force per unit length of wire, is given by

$$F_{\text{M}} = \mu_0 \frac{I_1 I_2}{2\pi d} \quad \text{(newton per metre)} \tag{5.11}$$

As in the case of ε_0, the constant μ_0 (*the permeability of free space* or *the magnetic constant*) is present to give the observed force when the currents are in amperes.

Maxwell in 1864 unified electricity and magnetism. He found that his theory predicted the existence of electromagnetic disturbances and waves which travelled in a vacuum with a speed v. In SI units, this speed is given by

$$v = 1/\sqrt{(\varepsilon_0\mu_0)},$$

$$\cong 3.0 \times 10^8 \text{ (metre per second)}.$$

(5.12)

Maxwell found that the number was close to the then known speed of light, and he therefore identified light as an electromagnetic wave.

At the end of the nineteenth century SI units did not exist and equations (5.10)–(5.12) could not be written in this form. At that time the ampere was defined as that steady current that would deposit a certain amount of silver in a period of one second in an agreed design of electrolytic cell. The coulomb was defined as the quantity of electric charge provided by one ampere in one second. The forces described by equations (5.10) and (5.11) could be measured and values of ε_0 and μ_0 (or at least their equivalents at that time) determined. These values, using the equivalent of equation (5.12), gave a value for the speed of light independent of a direct measurement of this quantity.

But there was a problem. It was expected that invariance under Galilean transformations would be true for electromagnetism as it was in classical mechanics for which *all inertial frames are equivalent*. Then, equations (5.10) and (5.11) had to be true in all inertial frames, with the same values for ε_0 and μ_0. Equation (5.12) in turn implied that the speed of light would be the same in all inertial frames, but a Galilean transformation requires that a velocity transforms as in equations (5.6)–(5.8). Therefore, in general, the speed of light should differ between different inertial frames. Thus electromagnetism appeared to be, and is, incompatible with Galilean relativity. An alternative statement is that Maxwell's electromagnetism was not *invariant* under Galilean transformations.

The favoured solution was to retain Galilean relativity for classical mechanics but to postulate the existence of a preferred inertial frame in which Maxwell's equations were correct and the speed of light was $1/\sqrt{(\varepsilon_0\mu_0)}$ (of course, a preferred frame was not good relativity!). In addition, nineteenth century physicists needed the idea of the existence of a medium to carry electromagnetic waves in the same way as a taut string can carry transverse waves. They called this medium the **ether** and it was at rest in the preferred inertial frame. This prompted a careful search for the effects of this ether. In Chapter 10 we shall return to this subject and describe some of the experiments performed in these endeavours. None found any convincing evidence for an ether. This was the serious situation which physicists such as Lorentz and Poincaré were trying to understand at the end of the nineteenth century.

Enter Einstein!

Before we yield the centre stage to Einstein, we draw our reader's attention to a brief description in Section 5.11 of the basis of the modern system of units. We have just used those units in a demonstration of the challenge that electromagnetism presented to Galilean relativity. Today the success of special relativity has allowed the values of c, μ_0 and $\varepsilon_0 = 1/\mu_0 c^2$ to be made fixed constants.

We recommend that at this time the reader make a relaxed diversion through Chapter 10. There we briefly describe some of the experiments performed to detect the ether and some of the attempts made to explain their failure.

**5.5   Einstein's relativity

Einstein solved the difficulties of reconciling electromagnetism and Galilean relativity with a new principle of relativity which later became known as *Special Relativity*. A modern form of his two postulates is:

1   All inertial frames are equivalent with respect to the laws of physics.
2   The speed of light in empty space is independent of the state of motion of its source.

Postulate 1 says that not only Newton's laws of motion but that all the laws of physics are the same in all inertial frames (that is *invariant*), gravity alone must be excluded. These laws had to include electromagnetism and Einstein was able to use this invariance of electromagnetism. Postulate 2 is not compatible with Galilean relativity; in that system, equation (5.8) tells us that the speed of light should depend on the speed of the source.

To find the consequences of his postulates Einstein had to abandon Newton's concept of a universal time and to find an alternative to equations (5.2) and (5.3) of the Galilean transformations. He presented this revolution and many important conclusions that flowed from it in his first relativity paper published in 1905. It has the rather intimidating title *On The Electrodynamics of Moving Bodies*.[2] In this chapter we shall map out Einstein's development of the subject and some subsequent advances up to the point that allows us to meet again the formulae for the energy and momentum of a particle.

The change away from Newton's universal time has consequences that cause the greatest difficulties to understanding special relativity. Einstein later gave an example, discussed in the next section, which illustrates the need for a change from our ingrained ideas about time.

**5.6   The rise and fall of simultaneity

As we have seen, Newton's concept of a universal time, a time that is the same in all frames (Section 5.3, Newton's assumption 1), means that two events found to occur at the same time by an observer stationary in one inertial frame will also

---

2   English translation in *The principle of relativity*. A collection of original memoirs on the special and general theory of relativity by H. A. Lorentz, A. Einstein, H. Minkowski and H. Weyl, New York, Dover Publications, Inc.: 1952.

be found to occur at the same time by an observer in any other inertial frame (equation (5.2)). There is no difficulty in accepting this as true for two events occurring simultaneously in front of the two observers. We could call that local simultaneity. Now consider two light-flash emitting events $\mathcal{P}$ and $\mathcal{Q}$ occurring at two separated points in space. It is natural to define these two events to be simultaneous if the light-flashes arrive at the same time at the point mid-way between the two space points. This is distant simultaneity.

However, this definition of distant simultaneity immediately brings in to play the propagation of light and, in particular, the finite speed of light. To see the consequences we consider Einstein's example. It is a *thought experiment* in that it cannot be realised as described, but serves to illustrate in familiar terms a possible situation. A railway train (Figure 5.2) is travelling with speed u along a straight section of the railway track. On the train, the guard is leaning out of a window that is half way down the length of the train. Standing beside the track is an inspector. We have two inertial frames: Σ is fixed to the ground on which the inspector stands and Σ' is fixed to the train. Consider, as would Newton, the following situation:

1   The guard is carried past the inspector and is opposite him at event $\mathcal{O}$.
2   Two events occur, $\mathcal{P}$ and $\mathcal{Q}$, both involving lightning strikes.
3   Both occur at the same time as does event $\mathcal{O}$, as determined by clocks in Σ, all of which are stationary with respect to the inspector.
4   The lightning of event $\mathcal{P}$ strikes the train nearer its leading end from the guard.
5   The lightning of event $\mathcal{Q}$ strikes the train nearer its trailing end from the guard so that the inspector is spatially mid-way between the two events $\mathcal{P}$ and $\mathcal{Q}$ (Figure 5.2).
6   Then it must be that the guard, when at event $\mathcal{O}$, is also spatially mid-way between the two events.

Figure 5.2 Einstein's train with the guard leaning out of a window. The inspector is standing beside the railway track watching the train pass. We catch the scene just as the guard is opposite the inspector. Lightning strikes the train at two points equidistant from the inspector and therefore simultaneously as observed by him. The guard also sees the two lightning flashes.

What are the consequences?

1   The inspector will see the two lightning flashes (using a suitable arrangement of mirrors) at the same instant and will judge the events $\mathcal{P}$ and $\mathcal{Q}$ to be simultaneous.

2   The finite speed of light means that this observation by the inspector will occur after event $\mathcal{O}$ by a time Δt, say.

3   During the time interval Δt the guard has been carried forward by the train. He does not remain opposite the inspector. Since he is carried towards the spatial position of event $\mathcal{P}$ it is reasonable to suppose that he will see event $\mathcal{P}$ before event $\mathcal{Q}$.

4   The finite speed of light means that the guard will not see the two events to be simultaneous. Thus simultaneity in one inertial frame can not be preserved in other frames.

Einstein used this example in his book on relativity, first published in 1916.[3] He needed to prepare each reader for the idea that there was no universal time and that events that appear simultaneous to one observer may not appear so to another. Only then could he go on to describe the nature of special relativity and some of its results.

This result also shows that the time interval between any two events need not be the same for observers stationed in different inertial frames.

In Section 7.7 we shall look at the status of simultaneity, or rather the lack of it, in special relativity.

**5.7   Time

So what did Einstein put in place of Newton's universal time? First, he considered a single inertial frame. He imagined each space point equipped with a stationary clock. An event at any particular point would be simultaneous with a unique reading on the clock stationed at this point. Einstein required that reading to be the time coordinate of the event.

> All this is, of course, an unrealisable abstraction. We cannot accommodate a clock at every space point or have a real clock and an event at the same point. However, we are all used to the idea of the instantaneous speed of a body, although it is a quantity that cannot be measured without taking a finite time Δt to do so. The instantaneous speed is found in an abstract way by considering what happens in the limit $\Delta t \rightarrow 0$. Similar steps can be followed in relativity by imagining clocks that become smaller and smaller, and closer and closer to the events. But that is enough metaphysics. Let us return to clocks.

Remember that all these clocks operate identically. But now we have the problem of synchronisation of all the clocks throughout the given inertial frame (which we

3 Einstein, A., *Relativity*, 15th edn, London: Methuen & Co., 1954.

label Σ). Here is a version of Einstein's instructions for the synchronisation of any clock with a clock at the origin of Σ.

1 Station a clock C_O at the origin O of the coordinates of Σ and another, C_P, at a point P.
2 Place at P a mirror facing the origin O.
3 Organise a device that launches a flash of light from the origin towards P where it is reflected back to the origin.
4 Record the times, t_1 and t_2, on the clock C_O at the departure from O and return to O of the flash of light.
5 Record also the time, t_P, on clock C_P of the event that is the reflection of the light flash at the mirror at P.
6 Adjust the clock C_P so that when the procedure is repeated the record shows that $t_P = (t_1 + t_2)/2$. The clock C_P is now synchronised with the clock C_O.
7 Repeat this procedure to bring all other clocks into synchronisation with C_O. Then all the clocks in the inertial frame are synchronised with one another.

Remember that was a prescription for the synchronisation of clocks in one inertial frame. If other inertial frames are involved, then the identical procedure must be used in each frame separately.

Note that these procedures assume that the speed of light is the same on the inward journey as it was on the outward journey. That property is implicit in Einstein's second postulate. We shall return to this matter in Section 8.2 in connection with the derivation of the Lorentz transformations.

We can do one thing to simplify the algebra without endangering Einstein's postulates. For our standard arrangement (Figure 5.1) of two inertial frames there is one event, $\mathbb{O}$, at which the origin of the Cartesian coordinates of Σ' coincides with the origin of the coordinates of Σ. It is usual to set the zero of time in both inertial frames so that at this event the time coordinates satisfy

$$(t)_{\mathbb{O}} = 0 = (t')_{\mathbb{O}}. \tag{5.13}$$

The subscript after the close of a bracket refers to the event. The brackets contain a coordinate of this event. A prime indicates a coordinate in Σ'.

Given this zeroing of clocks at event $\mathbb{O}$, consider a second event $\mathcal{P}$ that occurs at another space-time point. Suppose this event has coordinates t, x, y, z in Σ and coordinates t', x', y', z' in Σ'. Concentrating on the time we shall find that, even though the event $\mathbb{O}$ has $t = t'$, other events do not necessarily have $t = t'$. This is different from the Galilean result (equation (5.2), $t = t'$ always), and one which we anticipated in the discussion on simultaneity in the preceding section. It is this counter intuitive effect that causes many difficulties.

We can now make an addition to our list of terms that we started in Section 5.2. The **Lorentz Transformations** are the equations that connect quantities observed in one inertial frame with the same physical quantities observed in a second inertial frame. They satisfy Einstein's postulates. In Chapter 6 we shall learn how to transform the energy and momentum of a particle. In Chapter 7 we shall meet the Lorentz transformation equations that replace the Galilean equations (5.2) and (5.3). In Chapter 8 we shall derive these equations.

*5.8   Einstein's electrodynamics

For those interested but unable to read the originals in German, there is an English translation of Einstein's two relativity papers of 1905.[4] So what did Einstein do? He proposed and applied his postulates (Section 5.5) and in the ten sections of the first paper he laid the foundations of special relativity. Here we pick out from that paper the steps that lead to the formulae for the relativistic kinetic energy and momentum of a moving particle that we met in Chapter 2.

1   Einstein proposed the method for measuring the time coordinate of an event and described the procedure for the synchronisation of clocks throughout an inertial frame (Section 5.7).

2   He derived the equations for the Lorentz transformation of the coordinates of an event from an inertial frame Σ' to inertial frame Σ, as in Figure 5.1. We shall describe two derivations in Sections 8.3–8.5.

3   The relativity principle meant that Einstein could require Maxwell's equations to be covariant. That allowed him to determine the equations for the Lorentz transformation of the components of electric and magnetic fields between inertial frames Σ' and Σ. We shall derive those transformations in Sections 9.8 and 9.9. They show that electric and magnetic fields are not separate entities: thus an electromagnetic field which is observed to be solely electric in one inertial frame may be observed to have both electric and magnetic components in another inertial frame. He showed that the electric charge on a body was invariant.

4   Einstein derived the equations that transform a velocity between the inertial frames Σ' and Σ. We shall derive those equations in Section 7.15.

5   Einstein retained Newton's three laws of motion except that the second law, as in equation (5.1), could only be applied to a particle instantaneously at rest. By 1905 Lorentz had shown that the mass of an electrically charged particle moving through the ether would increase with its speed. In his investigations, Einstein must have anticipated a similar effect. Therefore if he was to use Newton's Second Law it could only be for a particle at rest, which is when he could be certain of its mass.

6   He considered the case of a particle, charge q, instantaneously at rest in an inertial frame Σ' (see Figure 5.1) and subjected to an electric field $\mathbf{E}'$. He applied Newton's Second Law to give the acceleration in that frame:

$$M \frac{d^2 \mathbf{r}'}{dt'^2} = q \mathbf{E}', \tag{5.14}$$

4 See footnote 2.

where $\mathbf{r}'$ is the vector representing the coordinates of the particle's position and M is its rest mass. The event at which the particle is momentarily at rest has a time coordinate t' in Σ'. Equation (5.14) is, of course, three equations:

$$M\frac{\mathrm{d}^2 x'}{\mathrm{d}t'^2} = qE'_{x'}, \quad M\frac{\mathrm{d}^2 y'}{\mathrm{d}t'^2} = qE'_{y'}, \quad M\frac{\mathrm{d}^2 z'}{\mathrm{d}t'^2} = qE'_{z'},$$

where x', y', z' are the coordinates of the charged particle at time t'.

7 He transformed this equation (5.14) from Σ' to the second inertial frame Σ of Figure 5.1. In this transformation the electric field $\mathbf{E}'$ observed in Σ' becomes an electric field $\mathbf{E}$ and a magnetic induction $\mathbf{B}$ when observed in Σ. The frame Σ' is moving with speed u in Σ in the z direction; therefore the particle instantaneously stationary in Σ' is observed in Σ to be moving with speed instantaneously u along the z-axis.

8 Einstein actually presented his procedure in component by component form and arrived at separate equations for each of the three components of the particle's acceleration in Σ at the instant it had speed u along the z-axis in the presence of electric and magnetic fields. These equations are

$$M\gamma_u \frac{\mathrm{d}^2 x}{\mathrm{d}t^2} = q(E_x - uB_y) = qE'_{x'}/\gamma_u,$$

$$M\gamma_u \frac{\mathrm{d}^2 y}{\mathrm{d}t^2} = q(E_y + uB_x) = qE'_{y'}/\gamma_u, \qquad (5.15)$$

$$M\gamma_u^3 \frac{\mathrm{d}^2 z}{\mathrm{d}t^2} = qE_z = qE'_{z'},$$

where $\gamma_u = 1/\sqrt{1 - u^2/c^2}$. Now unprimed quantities are those observed in Σ. We also show how the electric field $\mathbf{E}'$ in Σ' is observed to have electric and magnetic components in Σ. Transforming the acceleration terms in equation (5.14) is tricky and we do not do it here (see Problem 7.16). We derive the transformation of the components of the electromagnetic field in Section 9.9.

10 Following convention current at the time, Einstein discussed these three equations in terms of a longitudinal ($\gamma_u^3 M$) and a transverse mass ($\gamma_u^2 M$) for the accelerations along and perpendicular to the direction of u. However, shortly after Einstein's papers appeared, Planck gave a more transparent interpretation with a definition of the momentum of a particle. We have already met that definition in equation (2.3) and we shall discuss its connection with Einstein's result later in this section. Equations (5.15) do not need special attention from the reader. They are quoted to show what Einstein did and to give an example of what can happen in Lorentz transformations.

11 From his equations for the acceleration in Σ, Einstein showed that the same particle moving with a velocity $\mathbf{u}$ perpendicular to a magnetic induction

$\mathbf{B}(B = |\mathbf{B}|)$ moves in a circle with radius ρ given by

$$Bq\rho = \frac{Mu}{\sqrt{1 - u^2/c^2}}. \tag{5.16}$$

This is equation (4.5) introduced in Chapter 4.

12 By considering the particle accelerated through an electrostatic potential difference to a speed u, Einstein showed that the final kinetic energy T was given by

$$T = Mc^2 \left(\frac{1}{\sqrt{1 - u^2/c^2}} - 1 \right). \tag{5.17}$$

This is equation (2.4) introduced in Chapter 2.

13 Since $T \to \infty$ as $u \to c$, particles with non-zero rest mass cannot be accelerated to the speed of light. Thus there is a limiting speed, and it is the speed of light.

14 Einstein also deduced the equations for the Doppler effect and for the aberration of light. We shall examine these subjects in Sections 6.3 and 6.4. Then he derived the transformation equation (from Σ' to Σ) for the energy of a light wave moving in an arbitrary direction. In his second paper Einstein used this result to show that the mass of a body is a measure of its energy content and that if the energy content changes by L the mass changes in the same sense by L/c^2. Although the argument used by Einstein has been the subject of a critical debate,[5] the result is correct. Therefore, as we have described in Section 2.2, we can take the rest mass energy of a particle of mass M to be Mc^2.

Note that Einstein used the relative speed of our two inertial frames (for which we have used the symbol u) to derive this formula for the equation of motion of a particle having speed u. We intend to make a notational difference between the speed of a particle and the relative speed of two inertial frames. Normally we shall use v for the former and u for the latter.

**5.9   Comments on Einstein's results

The nub of Einstein's paper of 1905 is the derivation of the Lorentz transformation (between Σ and Σ') of the coordinates of an event from his two postulates. Earlier, Lorentz and others had known Maxwell's equations were covariant under this transformation but by training and experience they thought of the world system in terms that were based on material and mechanical ideas. For example: light propagates in a vacuum and in transparent materials, therefore there was a necessary and all pervading ether. Einstein freed physicists from these chains and introduced

---

5 Ives, H. E., *J. Opt. Soc. of America* **42**, 1952, 540–3. Stachel, J. and Torretti, R., *Amer. J. Phys.* **50**, 1982, 760–3.

the universal properties of space–time. We shall be concerned with the meaning and consequences of the Lorentz transformations in Chapters 6–9. In Chapter 8 we shall derive the transformation from Σ' to Σ. The remainder of this chapter is reserved for commenting on Einstein's electrodynamic results and their connection with the definition of relativistic momentum and energy.

Einstein's electrodynamic results are equations (5.15). These give the components of the acceleration of a charged particle in Σ which is momentarily at rest in Σ' and there subject to an electric field. If those equations are to be put in the form of Newton's Second Law, it is necessary to decide whether the factors which are powers of $\sqrt{1 - u^2/c^2}$ should be on the left- or right-hand side. Planck showed that the force $\mathbf{F}$ should be defined to be identical to the Lorentz force, equation (4.3). The result is that equations (5.15) are a particular case of a vector equation

$$M \frac{d}{dt} \left(\frac{\mathbf{v}}{\sqrt{1 - v^2/c^2}} \right) = q(\mathbf{E} + \mathbf{v} \times \mathbf{B}), \tag{5.18}$$

(equation 4.7) where $\mathbf{v}$ has components $v_x = 0$, $v_y = 0$, $v_z = u$. Problem 5.1 asks the reader to check this assertion. Then (5.18) is the equation of motion for a particle, charge q, rest mass M, and velocity $\mathbf{v}(v = |\mathbf{v}|)$ moving in an electric field $\mathbf{E}$ and magnetic induction $\mathbf{B}$. We can make contact with Newton's Second Law by using Planck's definition of the momentum of a particle:

$$\mathbf{P} = \frac{M\mathbf{v}}{\sqrt{1 - v^2/c^2}}, \tag{5.19}$$

instead of $\mathbf{P} = M\mathbf{v}$. Then equation (5.18) becomes

$$\frac{d\mathbf{P}}{dt} = \mathbf{F} = q(\mathbf{E} + \mathbf{v} \times \mathbf{B}). \tag{5.20}$$

We can also remind readers of the result that the energy of a particle is given by

$$E = Mc^2 + T = \frac{Mc^2}{\sqrt{1 - v^2/c^2}}. \tag{5.21}$$

The low velocity ($v \ll c$) limits for $\mathbf{P}$ in equation (5.19) and for T in equation (5.17) were given in Section 2.2 and are the same as the non-relativistic formulae for the same quantities.

The consequences of the points made are as follows:

1   Special relativity is consistent with Newton's laws of motion.
2   The formula for the Lorentz force is proved correct from the postulates of special relativity and Newton's laws.
3   The relativistic formulae for the kinetic energy and momentum of a particle are consistent with the pre-relativity formulae at low speeds.

Consider equation (5.20)

$$\mathbf{F} = \frac{d\mathbf{P}}{dt},$$

which now holds in special relativity if we accept the new definition of momentum. However, the factor $1/\sqrt{1 - v^2/c^2}$ present in the formula for $\mathbf{P}$, leads to a situation in which an acceleration generally needs a force with a component parallel to v in addition to that in the direction of the acceleration. See Problem 5.2.

Although not given by Einstein, equations (5.18)–(5.20) are consistent with his 1905 results. However, there was an important development in 1909. G. N. Lewis and R. C. Tolman showed that equations (5.19) and (5.21) for the momentum and energy of a single particle were consistent with the conservation of total momentum and total energy in particle collisions at relativistic speeds. We introduced relativistic momentum and energy for a particle in Chapter 2, and we assumed what Lewis and Tolman demonstrated when we discussed some relativistic kinematics in Chapter 3. We examine the method of Lewis and Tolman in Section 8.12.

**5.10   Charge invariance

Einstein showed that the charge carried by a particle is invariant (Section 5.8). In Chapter 4 we assumed this was the case but it is interesting to look briefly at the experimental evidence. Consider the residual charge on nominally electrically neutral atoms and molecules. This is measured to be less than about 10^{-19} of the charge on the electron. However, in such atoms the electrons are moving with speeds greater than $0.01c$ and the protons in the nucleus are moving with speeds about $0.3c$. Thus these speeds do not alter the overall charge significantly. Further evidence for charge invariance comes from the validity of equation (4.7), in particular, applied to particle accelerators that confine moving charged particles into near-circular closed orbits by using magnetic fields. Equation (4.7) correctly predicts the orbit curvature from non-relativistic speeds up to speeds that for protons have $\gamma (= 1/\sqrt{1 - v^2/c^2})$ equal to 1000 and for electrons that have γ equal to 2×10^5, without requiring a change in charge.

**5.11   Summary

We trust this chapter has allowed the reader to appreciate how electromagnetism challenged the Galilean principle of relativity. We have introduced Einstein's principle of relativity and outlined how he laid the foundations of special relativity. We also indicated that his results lead to the formulae for the conserved relativistic momentum and energy. We used those formulae freely in Chapters 2–4; in the next chapter we shall extend our discussion of relativistic kinematics to include the Lorentz transformation of momentum and energy.

In Section 5.4 we promised to give the up-to-date position on the electromagnetic aspect of SI units.

1   For time, the second is the period that elapses during a given number of periods of a certain transition in the ^{133}Cs isotope (Section 5.2).

2   The successful verification of special relativity leads us to be confident that the speed of light in a vacuum is a universal constant. So it can be given as an exact number (in Section 5.2, and represented by c). A metre of length is then the distance moved by light in $1/c$ seconds.

3   The value of the permeability of free space μ_0 is fixed to be $4\pi \times 10^{-7}$ Newton per (ampere)^2.

4   Since the speed of light is given by $c = 1/\sqrt{\varepsilon_0\mu_0}$, the permittivity of free space is given as $\varepsilon_0 = 1/(\mu_0 c^2)$ (Farad per metre).

5   Thus c, ε_0, and μ_0 are exact numbers.

6   One kilogram is the mass of a cylinder of platinum–iridium in Paris. From that and from the second and metre, the newton (force) can be fixed. From equation (5.10) the ampere can be determined. From the ampere follows the coulomb and so on.

That short summary does not reveal an immense and ongoing programme of very precise experimental work that is required to propagate these and other standards to all areas of quantitative science and engineering. Note the dependence on Einstein's second postulate.

Problems

5.1   Consider the equation

$$M\frac{d}{dt}\left(\frac{\mathbf{v}}{\sqrt{1 - v^2/c^2}}\right) = q(\mathbf{E} + \mathbf{v} \times \mathbf{B})$$

for the special case that $\mathbf{v}$ has components $v_x = 0$, $v_y = 0$, $v_z = u$. Show that it reduces to the equations

$$M\gamma_u \frac{d^2x}{dt^2} = q(E_x - uB_y),$$

$$M\gamma_u \frac{d^2y}{dt^2} = q(E_y + uB_x),$$

$$M\gamma_u^3 \frac{d^2z}{dt^2} = qE_z,$$

where $\gamma_u = 1/\sqrt{1 - u^2/c^2}$.

5.2   Consider the equation

$$\frac{d\mathbf{P}}{dt} = \mathbf{F} \quad \text{where } \mathbf{P} = \frac{M\mathbf{v}}{\sqrt{1 - v^2/c^2}}.$$

Show that to produce an acceleration of a particle in a particular direction requires a force $\mathbf{F}$ not solely in the direction of the acceleration but with a component along the direction of the particle's existing velocity.

6   Relativistic kinematics II

**6.1   Introduction

The next step in our confidence building programme is to examine the Lorentz transformation of the energy and momentum of a particle. These transformations do just what might be expected. The momentum of a particle is observed to have a greater z component in Σ than z' component in Σ' (Figure 6.1). This is the same effect as would be observed in a corresponding Galilean transformation where it is easily understood. The quantitative details are different.

These transformations are important in elementary particle physics and in astrophysics. For example, in the case of particle collisions the most important transformations are those between a centre-of-mass, a particle rest frame, and the laboratory. The reason for this is that many particle collisions are often most readily analysed kinematically and dynamically in a centre-of-mass frame, but are observed in a laboratory frame. A similar situation applies to particle decays. In astrophysics, many violent events have sources moving at relativistic velocities relative to the Earth but their electromagnetic radiation is observed in an Earth-bound reference frame. The interpretation of observational data involves Lorentz transformations.

We shall present the Lorentz transformation for energy and momentum but leave its justification to Section 8.10. In this chapter we shall describe situations in optical and particle physics where the transformation is involved.

**6.2   The Lorentz transformation of energy and momentum

We bring into play our standard arrangement of two inertial frames: see Section 5.2 and Figure 5.1. In Figure 6.1 we have sketched the coordinate axes for the frame Σ' and Σ well separated. This is to enable us to show how the momentum vector of a particle observed to be $\mathbf{P}'$ in Σ' might be observed in Σ: the same particle will be observed to have a different momentum $\mathbf{P}$. Again we use u to represent the speed of Σ' in Σ.

A Lorentz transformation of the kind that we might make between inertial frames Σ and Σ' in Figure 6.1 is sometimes called a **Lorentz boost**. This is a handy piece of jargon since it distinguishes a simple boost from what could be a more general

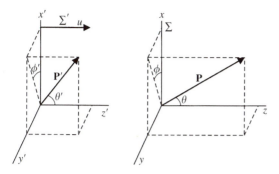

Figure 6.1 This is a modified version of Figure 5.1. Inertial frame Σ' is moving with speed u in frame Σ. The orientations of the two Cartesian coordinate systems are as in the standard arrangement of Figure 5.1. A moving particle is observed from Σ' to have energy E' and momentum $\mathbf{P}'$. The same particle is observed from Σ to have energy E and momentum $\mathbf{P}$.

As drawn we have

$$P'_{x'} = P'\sin\theta'\cos\phi', \qquad P_x = P\sin\theta\cos\phi,$$
$$P'_{y'} = P'\sin\theta'\sin\phi', \qquad P_y = P\sin\theta\sin\phi,$$
$$P'_{z'} = P'\cos\theta', \qquad P_z = P\cos\theta.$$

A result of the Lorentz boost along the z-axis is already anticipated in this drawing by attempting to show on a paper plane that in three dimensions

$$\phi' = \phi, \quad P'_{x'} = P_x \quad \text{and} \quad P'_{y'} = P_y.$$

and more complicated transformation involving rotations and displacements (see Section 8.6).

We can start from the more familiar Galilean transformation. Equations (5.6)–(5.8) give the velocity transformations. Multiply each throughout by the particle's mass M and we obtain the transformations for Newtonian momentum:

$$P_x = P'_{x'},$$
$$P_y = P'_{y'}, \tag{6.1}$$
$$P_z = P'_{z'} + Mu.$$

The transformation for kinetic energy is then

$$T = T' + \tfrac{1}{2}Mu^2 + uP'_{z'}. \tag{6.2}$$

We now write down the Lorentz transformation for energy and momentum. They are

$$E = \gamma_u(E' + u P'_{z'}),$$
$$P_x = P'_{x'},$$
$$P_y = P'_{y'},$$
$$P_z = \gamma_u(P'_{z'} + \{u E'\}/c^2),$$

(6.3)

where

$$\gamma_u = 1/\sqrt{1 - u^2/c^2}.$$

This is not the most memorable form in which to write the transformation equations. We shall present a simpler form after checking that equations (6.3) for this Lorentz boost are acting as we expect, if necessary with the help of Figure 6.1. There are five points to consider:

1   We leave to the reader to verify (Problem 6.1) that the transformation satisfies the invariance

$$E'^2 - \mathbf{P}' \cdot \mathbf{P}' c^2 = E^2 - \mathbf{P} \cdot \mathbf{P} c^2.$$

We recall that this quantity is the square of the particle's rest mass energy, which must be an invariant (equation (2.6)).

2   The reciprocal transformation, that is from Σ to Σ' and transforming $E, \mathbf{P}$ to $E', \mathbf{P}'$ is given by the interchanges $E \leftrightarrow E', \mathbf{P} \leftrightarrow \mathbf{P}'$, and the change $u \to -u$ in equations (6.3). Applying the appropriate transformation from Σ' to Σ and then back again to Σ' must give the original values for the energy–momentum. We ask the reader to check this (Problem 6.2).

3   In the non-relativistic limit the Lorentz equations must become the Galilean equations. Thus in this limit we have

$$u/c \to 0, \quad v/c \to 0, \quad v'/c \to 0,$$

where v (v') stand for the particle's speed in Σ (Σ'). Then, for the kinematic variables

$$\mathbf{P} \to M\mathbf{v} \quad \text{and} \quad T \to \tfrac{1}{2}Mv^2$$

in Σ (similarly in Σ'). In addition, using the binomial expansion, we have

$$\gamma_u = 1/\sqrt{1 - u^2/c^2} = 1 + \tfrac{1}{2}(u^2/c^2) + O(u^4/c^4),$$
$$\gamma_u E' = \gamma_u(T' + Mc^2)$$
$$= Mc^2 + \tfrac{1}{2}Mu^2 + Mc^2 O(u^4/c^4) + T'\big(1 + \tfrac{1}{2}(u^2/c^2) + O(u^4/c^4)\big),$$

where $O(u^4/c^4)$ means terms of the order of u^4/c^4 and higher. Then, in equations (6.3), neglecting terms quadratic or of higher power in u/c:

$$E = \gamma_u(E' + u P'_{z'}) \qquad \rightarrow \qquad T = T' + \tfrac{1}{2}M u^2 + u P'_{z'},$$
$$P_z = \gamma_u(P'_{z'} + \{u E'\}/c^2) \quad \rightarrow \quad P_z = P'_{z'} + M u.$$

Thus we have, as required, regained the Galilean transformations (equations (6.1) and (6.2)).

4   For the relative motion shown in Figure 6.1, we expect that a particle moving in Σ' will be observed to have the z component of its momentum in Σ greater than that observed in Σ'. That is what the fourth equation of (6.3) requires since for our situation $u > 0$ and $\gamma_u > 1$.

5   If the particle is at rest in Σ' ($E' = M c^2$, $\mathbf{P}' = 0$) the Lorentz boost gives

$$E = \gamma_u M c^2, \quad P_x = 0, \quad P_y = 0, \quad \text{and} \quad P_z = \gamma_u M u.$$

In Σ we expect that the particle will appear to be moving with velocity u in the direction of the z axis. If we replace v with u in the equations in Table 3.1 then we obtain these values for E and P.

Thus our kinematic and transformation equations are self-consistent, have the correct non-relativistic limit and preserve the mandatory invariance. We postpone to Section 8.10 the justification of these Lorentz transformations for energy and momentum. For the present, we can discuss some applications with confidence that we are on firm ground.

The equations of the Lorentz boost look complicated. The reason is that they mix E and P which have different dimensions. Therefore we can simplify by rearranging so that each $\mathbf{P}$ is a $\mathbf{P}c$ and using

$$\beta_u = u/c,$$

which means that

$$\gamma_u = 1/\sqrt{1 - \beta_u^2}.$$

We are putting subscript's u on β and γ to remind us that these are the parameters of the transformation, not any fractional speed or any Lorentz factor of the moving particle. Then equations (6.3) become

$$
\begin{aligned}
E &= \gamma_u(E' + \beta_u P'_{z'} c), \\
P_x c &= P'_{x'} c, \\
P_y c &= P'_{y'} c, \\
P_z c &= \gamma_u(P'_{z'} c + \beta_u E').
\end{aligned}
\tag{6.4}
$$

These are simpler to remember than equations (6.3) but the reader may find the following matrix equation is even simpler:

$$\begin{pmatrix} E \\ P_x c \\ P_y c \\ P_z c \end{pmatrix} = \begin{pmatrix} \gamma_u & 0 & 0 & \gamma_u \beta_u \\ 0 & 1 & 0 & 0 \\ 0 & 0 & 1 & 0 \\ \gamma_u \beta_u & 0 & 0 & \gamma_u \end{pmatrix} \begin{pmatrix} E' \\ P'_x c \\ P'_y c \\ P'_z c \end{pmatrix}. \tag{6.5}$$

(Of course, all would be even simpler if we changed $Pc \rightarrow P$, that is, use units in which the speed of light is 1. We shall not do that and thereby stick to the policy decided in Section 3.1.)

The square matrix in equation (6.5) we call $\mathbf{L}(\beta_u)$. The reciprocal transformation $(\Sigma \rightarrow \Sigma')$ is performed with a matrix $\mathbf{L}^{-1}(\beta_u)$. From point 2 above we know that

$$\mathbf{L}^{-1}(\beta_u) = \mathbf{L}(-\beta_u).$$

Also the successive boosts $\Sigma' \rightarrow \Sigma \rightarrow \Sigma'$ restore the original values for the components of the energy–momentum. Therefore the square matrix $\mathbf{L}$ satisfies

$$\mathbf{L}(\beta_u)\mathbf{L}(-\beta_u) = \mathbf{I} \quad \text{(the } 4 \times 4 \text{ unit matrix).}$$

See Problem (6.3). Figure 6.2 summarises the formulae for a Lorentz boost for a momentum in the xz plane. We are ready to discuss some applications.

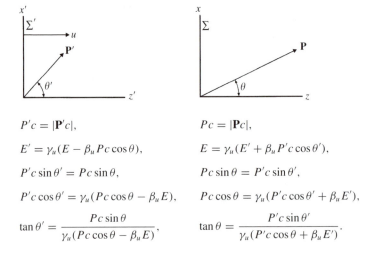

$$P'c = |\mathbf{P}'c|, \qquad\qquad Pc = |\mathbf{P}c|,$$

$$E' = \gamma_u(E - \beta_u P c \cos \theta), \qquad E = \gamma_u(E' + \beta_u P'c \cos \theta'),$$

$$P'c \sin \theta' = Pc \sin \theta, \qquad Pc \sin \theta = P'c \sin \theta',$$

$$P'c \cos \theta' = \gamma_u(Pc \cos \theta - \beta_u E), \qquad Pc \cos \theta = \gamma_u(P'c \cos \theta' + \beta_u E'),$$

$$\tan \theta' = \frac{Pc \sin \theta}{\gamma_u(Pc \cos \theta - \beta_u E)}, \qquad \tan \theta = \frac{P'c \sin \theta'}{\gamma_u(P'c \cos \theta' + \beta_u E')}.$$

Figure 6.2 This figure summarises the equations for the Lorentz boosts for the energy and momentum of a single particle moving in the $x'z'(xz)$ plane in $\Sigma'(\Sigma)$. The same equations apply to the total energy and momentum of a group of particles with its centre-of-mass moving in the same plane.

*6.3 The Lorentz boost for optics: aberration

Light is photons. Photons behave as do particles of zero rest mass. Therefore we can use the equations for the transformation of energy and momentum of photons to obtain relations between the frequency and direction of light in different inertial frames.

We shall deal with a source that emits, in its rest frame, light of frequency f_0. (Its *proper frequency*.) Thus we are to deal with the photons as if they are particles with energy hf_0 and momentum hf_0/c in the rest frame of the source, where h is Planck's constant. We need to be able to find the frequency and direction of such photons in an inertial frame in which the source is moving. To this end we use the standard arrangement shown in Figure 6.1. Station the source at the origin of Σ' and consider photons emitted at a polar angle θ', and an azimuthal angle $\phi' = 0$, that is in the $x'z'$ plane. (All the essential physics may be found if the photon is emitted in this plane and including $\phi' \neq 0$ is an easy but unnecessary complication.) So Σ' is the frame in which the source is at rest. The photons have energy–momentum given by $(E', \mathbf{P}'c) = (hf_0, hf_0 \sin\theta', 0, hf_0 \cos\theta')$. Transform the energy–momentum to Σ (see Figure 6.2):

$$
\begin{pmatrix} E \\ P_x c \\ P_y c \\ P_z c \end{pmatrix} = \begin{pmatrix} \gamma_u & 0 & 0 & \gamma_u \beta_u \\ 0 & 1 & 0 & 0 \\ 0 & 0 & 1 & 0 \\ \gamma_u \beta_u & 0 & 0 & \gamma_u \end{pmatrix} \begin{pmatrix} hf_0 \\ hf_0 \sin\theta' \\ 0 \\ hf_0 \cos\theta' \end{pmatrix}.
\tag{6.6}
$$

These particular photons will be observed with frequency f at a polar angle θ, ϕ so we can replace the left hand side and evaluate the right-hand side of this equation to obtain

$$
\begin{pmatrix} hf \\ hf \sin\theta \cos\phi \\ hf \sin\theta \sin\phi \\ hf \cos\theta \end{pmatrix} = \begin{pmatrix} \gamma_u hf_0(1 + \beta_u \cos\theta') \\ hf_0 \sin\theta' \\ 0 \\ \gamma_u hf_0(\cos\theta' + \beta_u) \end{pmatrix}.
\tag{6.7}
$$

Thus

$$
\phi = 0,
\tag{6.8}
$$

$$
f = f_0 \gamma_u(1 + \beta_u \cos\theta'),
\tag{6.9}
$$

$$
\cos\theta = \frac{\cos\theta' + \beta_u}{1 + \beta_u \cos\theta'}.
\tag{6.10}
$$

A useful alternative to equation (6.10) is

$$
\tan\theta/2 = \sqrt{\frac{1 - \beta_u}{1 + \beta_u}} \tan\theta'/2.
\tag{6.11}
$$

These equations describe the transformation of photons from Σ' to Σ. We note some properties:

1   The $\phi = 0$ solution for equation (6.7) follows from the invariance of the transverse component of the momentum in the transformation.
2   Equation (6.9) applies to photons of any frequency. Since it is unusual to refer to the frequency of high energy photons we can easily multiply each side by h (Planck's constant) and it is then correct for photon energy.
3   Equation (6.10) is remarkable. It relates $\cos\theta$ to $\cos\theta'$ and that relation depends on the speed of Σ' in Σ. However, it is completely independent of the photon energy. This is remarkable because the corresponding equations for the transformation for the energy–momentum of particles with rest mass give a relation between polar angles that does depend on the particle energy.
4   The observed difference in the direction of propagation of light as between two different inertial frames is called **aberration**. Equations (6.10) and (6.11) are the relativistic formulae for this effect.
5   Note that equation (6.9) gives the photon frequency observed in Σ in terms of the angle of emission with respect to the z'-axis in Σ'. If we solve equation (6.10) for $\cos\theta'$ in terms of $\cos\theta$ we obtain

$$\cos\theta' = \frac{\cos\theta - \beta_u}{1 - \beta_u \cos\theta}. \tag{6.12}$$

Substituting this into equation (6.9) gives

$$f = f_0 \frac{1}{\gamma_u(1 - \beta_u \cos\theta)}. \tag{6.13}$$

This equation gives the photon frequency in Σ in terms of the source frequency and the angle θ in Σ between the direction of the detected photon and the velocity of the source before emission of the photon.

The basic results are summarised in Figure 6.3. Note particularly that the results apply to light emitted by a source in Σ' (or in Σ) and observed in Σ (or in Σ').

There is one aspect of aberration that can be a source of confusion. Light is often observed by eye, either directly or through a telescope, for example. The line-of-sight, or the direction of pointing of the telescope, is directly opposite to that of propagation of the detected photons. The formulae derived and given in Figure 6.3 contain angles defined by the direction of propagation with respect to the boost, not by the the line-of-sight.

An example of these difficulties concerns a source (frequency f_0) moving with velocity **u** in Σ but not directly away from or towards an observer stationed in Σ. A rest frame of the source is Σ': Lorentz boost along **u** into a rest frame of the observer. The observer receives light from along his line-of-sight and measures frequency f. Suppose this observer knows θ', the angle in Σ' between **u** and the direction of photons heading towards him or her. It is equation (6.9) that will allow

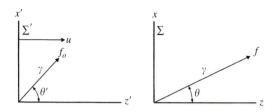

Aberration of light emitted from a source (frequency f in Σ or f_0 in Σ'):

$$f_0 = f\gamma_u(1 - \beta_u \cos\theta) \qquad\qquad f = f_0\gamma_u(1 + \beta_u \cos\theta')$$

$$= \frac{f}{\gamma_u(1 + \beta_u \cos\theta')} \qquad\qquad = \frac{f_0}{\gamma_u(1 - \beta_u \cos\theta)}$$

$$\cos\theta' = \frac{\cos\theta - \beta_u}{1 - \beta_u \cos\theta} \qquad\qquad \cos\theta = \frac{\cos\theta' + \beta_u}{1 + \beta_u \cos\theta'}$$

NB: All angles refer to the direction of propagation of the photons with respect to the z- or z'-axis. See the end of Section 6.3.

The longitudinal Doppler effect:

If $\theta' = 0°$ **then $\theta = 0°$**

and

$$f_0 = f\sqrt{\frac{1 - \beta_u}{1 + \beta_u}} \qquad\qquad f = f_0\sqrt{\frac{1 + \beta_u}{1 - \beta_u}}$$

The transverse Doppler effect:

If $\theta = 90°$

then

$$f_0 = f\gamma_u \qquad\qquad f = f_0/\gamma_u = f_0\sqrt{1 - \beta_u^2}$$

$$\cos\theta' = -\beta_u \qquad\qquad \cos\theta = 0$$

Figure 6.3 This figure summarises the formulae for the aberration of light emitted from a source and for the Doppler effect. Of course, the formulae are not restricted to optics but apply to the whole electromagnetic spectrum and to any particles of zero rest mass.

a calculation of f_0 from f, or the other way around. Alternatively, if θ is known, it is equation (6.13) that must be used.

The reader must be mindful of these difficulties in using the formulae of aberration.

We have derived these formulae assuming that photons behave as do particles. The identical results may be obtained from the wave model of light. We give an example of that in Section 9.3.

*6.4 The relativistic Doppler effect

Doppler was the first to record the effect in which the pitch of a sound is apparently raised or lowered as the source approaches or recedes from the listener. The similar effect was later found for light sources. Pre-Einstein it was thought that, for an observer at rest with respect to the ether, the observed frequency f would be $f_0/(1 \pm u/c)$ for a source moving towards $(-)$ or away $(+)$ from the observer with speed u through the ether. Special relativity provided the correct formula.

Of course, the Doppler effect for the general angle of observation is given by equation (6.9). However, there are particular cases, longitudinal and transverse.

a. *The longitudinal Doppler effect.* Consider a source of photons moving with speed u directly towards a particular observer. This corresponds to the situation in the previous section in which the source is stationary in Σ' and emits light at an angle with $\theta' = 0$. This light is detected by an observer stationary in Σ. Then $\cos \theta' = 1$ in equations (6.6)–(6.12). The result is that this observer 'sees' photons with a frequency f_+ given by

$$f_+ = \gamma_u(1 + \beta_u)f_0 = f_0\sqrt{\frac{1 + \beta_u}{1 - \beta_u}}. \tag{6.14}$$

It is obvious that for the source moving directly away from this observer the shifted frequency is given by

$$f_- = f_0\sqrt{\frac{1 - \beta_u}{1 + \beta_u}}. \tag{6.15}$$

Note that

$$f_0 = \sqrt{f_+ f_-}. \tag{6.16}$$

For source speeds that are not relativistic ($u^2 \ll c^2$) equations (6.14) and (6.15) reduce to the pre-relativity result (for an observer at rest in the ether!):

$$f_\pm = f_0/(1 \mp u/c).$$

b. *The transverse Doppler effect.* Let us ask a question. It concerns a particular observer stationed on the x-axis of the inertial frame Σ. This observer has arranged to measure the frequency of light received from a source at the origin O of Σ. If the source is at rest at O then the frequency measured will be f_0. What frequency will be observed if the source is moving with speed u along the z-axis as it arrives at the origin and emits a photon? There is a downward shift in the frequency observed. This is a purely relativistic effect. It is called the *transverse Doppler effect* because it concerns a source that crosses the line-of-sight of a particular observer at right angles as measured in that observer's rest frame. Instantaneously there is no relative radial motion. Therefore equation (6.13) applies with $\theta = 90°$ giving

$$f = f_0/\gamma_u = f_0\sqrt{1 - \beta_u^2}.$$

It follows that $f < f_0$.

It is interesting to note one thing about the aberration involved in the transverse Doppler effect. Basing the discussion on the standard arrangement, we see that the photon is detected in Σ at 90° to the direction of the motion of the source along the z-axis. In Σ', a rest frame of the source, the photon is emitted at an angle θ' given by equation (6.12) with $\cos\theta = 0°$. Therefore

$$\cos\theta' = -\beta_u.$$

This means that the photon has to be emitted at an angle with respect to the z'-axis in Σ' greater than 90° in order that the aberration places it at 90° in Σ.

The formulae for these particular Doppler effects have been added to Figure 6.3. In this chapter we have chosen to obtain them from the energy–momentum transformations but they may also be found using the properties of time dilation. We deal with this effect in the next chapter.

*6.5 Aberration and the Doppler effect in astrophysics

a. *Stellar aberration.* The Earth is moving at a speed close to 30 km s^{-1} in its near-circular orbit around the Sun. An Earth-bound astronomer measures the north–south angular position of a star that is close to his zenith once a day. The astronomer will find over a complete year, that this angle varies ±20.5 arcsec. (1 arcsec = 1°/3600.) A simple illustration and calculation of this effect is shown in Figure 6.4.

This phenomenon was discovered and explained by Bradley nearly two centuries before Einstein.[1] Relativity is not essential to understanding the effect or for its quantitative evaluation. However, the existence of stellar aberration is an important fact in arguments made against the existence of an ether. We return to this matter in Chapter 10.

b. *Photon source properties in astrophysics.* The observation of line spectra in emission or in absorption permits the identification of atomic and molecular sources and absorbers. Observation of astronomical sources (and absorbers) may

---

1 Stewart, A. B., The discovery of stellar aberration, *Scientific American* **210**(3), 1964, 100–9.

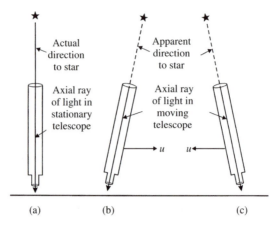

Figure 6.4 A simple illustration of stellar aberration. Diagram (a) shows a telescope set to observe an overhead star as if the Earth was stationary with respect to the fixed stars. If the observing astronomer wishes to keep the stellar image on the axis of the telescope he must compensate for the Earth's movement in its orbit around the Sun. This means tilting the telescope axis to be parallel to the vector difference of the light velocity and the Earth's velocity, as in (b). Six months later, the velocity of the earth has become reversed and the telescope must be tilted in the opposite direction, as in (c).

- The tilt is u/c radians, where u is the Earth's speed in orbit, which is close to $29.8\,\mathrm{km\,s^{-1}}$. Then the tilt is 9.935×10^{-5} radians or 20.5 arcsec. This is not a relativistic calculation. The reader is invited to use equation (6.10) to show that the relativistic result is negligibly different.
- Of course, the 'fixed' stars are not fixed and the Solar system is moving. The effect of these movements is very small on the scale of six months and do not affect the argument given above.

yield line frequencies that are broadened and shifted. Nonetheless the pattern of lines and their fine structure frequently allow source identification. Then the line broadening will give information about the source temperature. A shift from the identified and therefore known source frequency f_0 to the observed frequency f will give some information on the source velocity. This is an example of a difficult situation discussed at the end of Section 6.3. We indicated that equation (6.13) related f, f_0, the source speed $u(=\beta_u c)$, and the angle θ between the direction of photon propagation and the source velocity, as defined in the observer's rest frame. The result is

$$\gamma_u(1 - \beta_u \cos\theta) = f_0/f.$$

For uniform source speeds this equation will do no more than limit θ and u to certain ranges of linked values. Other information would have to be found to obtain separated values. Another kind of source is one that is in orbit around the centre of gravity of a binary star system. Now the Doppler shift varies around or near the frequency f_0 as the source traverses its orbit. Such variation will depend upon the period, the speed, the eccentricity, and the inclination of the plane of the orbit to the line-of-sight, and upon the velocity (relative to the observer) of the double system as a whole.

Problems 6.5–6.11 concern the Doppler effect and aberration in astrophysics as well as in other circumstances.

6.6 Lorentz boosts and new variables

Equation (6.5) is correct but there is another way of dealing with Lorentz boosts for energy and momentum. We start this discussion in the environment of Σ and Σ' as in Figure 6.1 with, as usual, the boost along the z-axis. We assume the freedom to rotate the coordinate axes around the $z(z')$ direction so that the component of momentum transverse to this direction is in the y direction. We do not lose any physics insight from this simplification. (Of course, there will be situations in which there may be several particles and any analysis may rely on keeping track of all three components properly in inertial frames not oriented for simplicity.)

Given a single particle of mass M moving in the yz plane of Σ', we rename P'_y the transverse momentum P'_T, 'transverse' meaning at a right angle (in Σ') to the boost direction. P'_z we rename the longitudinal momentum P'_L for being along the boost direction. And of course in Σ we shall have likewise P_T and P_L. We now extract from equation (6.4) those that apply to the transformation:

$$
\begin{aligned}
E &= \gamma_u(E' + \beta_u P'_L c), \\
P_T c &= P'_T c, \\
P_L c &= \gamma_u(P'_L c + \beta_u E').
\end{aligned} \tag{6.17}
$$

Here

$$
E = \sqrt{(Mc^2)^2 + (P_T c)^2 + (P_L c)^2}, \tag{6.18}
$$

and likewise for the primed quantities.

The reader may prefer to have equations (6.17) in matrix form:

$$
\begin{pmatrix} E \\ P_T c \\ P_L c \end{pmatrix} = \begin{pmatrix} \gamma_u & 0 & \gamma_u \beta_u \\ 0 & 1 & 0 \\ \gamma_u \beta_u & 0 & \gamma_u \end{pmatrix} \begin{pmatrix} E' \\ P'_T c \\ P'_L c \end{pmatrix}. \tag{6.19}
$$

The 3×3 matrix is just $\mathbf{L}(\beta_u)$ with the second row and the second column removed (Section 6.2). The angles between each momentum and the direction of boost are given by:

$$\tan \theta' = \frac{P_T'}{P_L'}, \quad \tan \theta = \frac{P_T}{P_L} = \frac{P_T' c}{\gamma_u (P_L' c + \beta_u E')}. \tag{6.20}$$

Throughout equations (6.17)–(6.20), the reciprocal relations may be found by adding primes to unprimed E, P and θ, removing primes on E', P' and θ', and changing β_u to $-\beta_u$.

The advantage of this approach is that we can set aside for the present all long-winded talk of vector components in Cartesian coordinates and in inertial frames Σ and Σ'. There is no mention of x, or of y, or of z, in equations (6.17)–(6.20). The boost gives us the sole direction required to make the transformation. The only vestige of Σ and of Σ' that remains is the primed and unprimed variables. The former refer to quantities in the frame that is to be boosted, the latter to the same quantities in the target frame, or the other way round!

There is another variable that is a useful alternative to P_L. It is called *rapidity*, usual symbol y. It is defined (in Σ) by

$$y = \frac{1}{2} \ln \left(\frac{E + P_L c}{E - P_L c} \right), \tag{6.21}$$

and similarly by y' in Σ'. The relation under the usual boost is

$$y = y' + \tanh^{-1} \beta_u. \tag{6.22}$$

This means that in a Lorentz boost the rapidity changes by a constant independent of its value in any frame. And, if there are two particles involved, the difference between their rapidities is unchanged in a boost. (Try Problem 6.12.)

What is the use of such variables? Look at Figure 6.5. This shows a bubble chamber photograph of the trajectories of 24 GeV/c protons traversing liquid hydrogen. One proton has collided with a stationary proton giving rise to 14 charged secondary particles and an unknown number of neutral particles. The collision of a second proton has produced four charged secondaries. Averaged over many of these inelastic proton–proton collisions the following facts emerge. If N is the average number of secondary particles observed in the laboratory frame then:

1   The distribution of secondaries in longitudinal momentum satisfies

$$\frac{dN}{dy} \cong \text{constant}$$

for $1 < y < 2$ with a turn-on from 0 to the constant in an interval $\Delta y \cong 1$ at either end of the constant region. The constant region expands logarithmically with increasing centre-of-mass collision energy. The turn-on intervals and the 'constant' change only slowly.

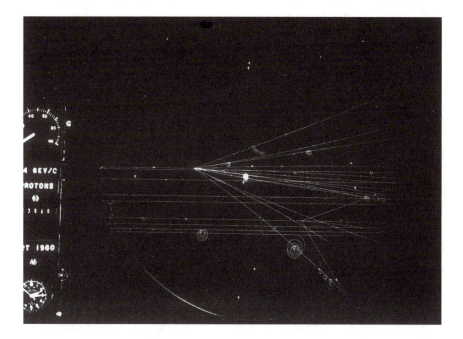

Figure 6.5 A bubble chamber photograph of the trajectories of 24 GeV/*c* protons in liquid hydrogen. Nine protons entered the chamber from the left, of which two suffered inelastic collisions. The more spectacular had 14 charged secondary particles and an unknown number of neutral secondaries. Another collision produced four charged secondaries. The centre-of-mass energy was 6.84 GeV in these collisions. This was sufficient to provide the mass and kinetic energy of the two nucleons (generic for neutron and proton) which had to be among the final particles, as well as a number of pions and possibly other kinds of particle.

The dimensions of the chamber are such that the distance across the visible field was about 30 cms. There was a magnetic induction of 1.5 tesla oriented as if perpendicular to, and into, the plane of the photograph.

2   The transverse momentum is defined with respect to the direction of the incident particle. The distribution of the transverse momentum of the secondary particles is given by

$$\frac{\mathrm{d}N}{\mathrm{d}P_\mathrm{T}^2} \propto \exp(-\alpha P_\mathrm{T}),$$

with $\alpha \approx 6 \ (\mathrm{GeV}/c)^{-1}$ for $P_\mathrm{T} < 1 \ \mathrm{GeV}/c$, which includes almost all the secondaries. The parameter α is almost independent of y. As the collision energy increases more particles are observed at large P_T than expected from this exponential fall.

The collisions to which these properties apply are those that involve two hadrons. These are the particles believed to be built of quarks. The properties have been described rather inexactly as there are variations from a complete universality in such collisions. However, the similarities are much greater than the differences, hence the interest in the variables y and P_T. The similarities must reflect a common mechanism in these inelastic collisions. The theory which attempts to describe such a mechanism is quantum chromodynamics (see Section 10.7).

$^+$6.7   Graphical representation of the Lorentz boost

There is a graphical representation of the transformation of momentum from one inertial frame to another. The construction is shown in Figure 6.6, and is demonstrated in the context of a direction of boost and of the longitudinal and transverse variables. Thus, in this figure, suppose $O'B_1$ is a measure of the momentum of a particle moving at an angle θ_1' to the direction of the boost (z'-axis) in an inertial frame Σ'. Boost to a second frame Σ, then OC_1 is the particle's momentum and direction in that frame.

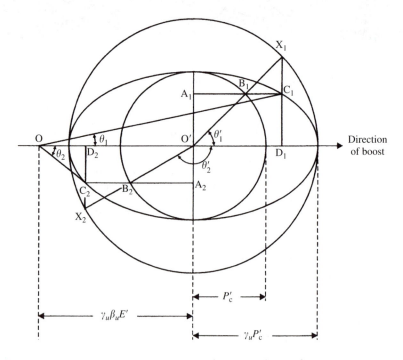

Figure 6.6 The graphical representation of a Lorentz boost of energy–momentum.
(Continued)

$^+$ This is the first use of this symbol to suggest that the material of a section is more suitable for the more advanced students.

Figure 6.6 (Continued)

1. The boost is made with a speed u from Σ' to Σ as in Figure 6.1. As usual $\beta_u = u/c$ and $\gamma_u^{-1} = \sqrt{1 - \beta_u^2}$. The energy–momentum $(E', \mathbf{P}'c)$ to be boosted is that of a particle in Σ'. The boost gives the energy–momentum $(E, \mathbf{P}c)$ of the same particle in Σ.

2. The construction transforms the momentum of a particle. The energies must be calculated from the momenta and the mass. A right angle triangle may be used to find E from Pc and Mc^2. All momenta are in the plane of this diagram.

3. The smaller circle has a radius $P'c$, centre O'. This sets the scale of the diagram. The larger circle has radius $\gamma_u P'c$ and centre O'. The ellipse (centre O') has its semi-major axis in the direction of the boost. It has length $\gamma_u P'c$. The semi-minor axis has length $P'c$. The construction of points on the ellipse is given in 5.

4. $O'B_1$ represents the momentum $\mathbf{P}'c$ of the particle when it is in Σ' moving at an angle θ_1' with respect to the boost direction. Then $P_{T}c = P'c \sin \theta_1' = O'A_1$ and $P_L'c = P'c \cos \theta_1' = A_1 B_1$, where $A_1 B_1$ is parallel to the boost direction.

5. Extend $O'B_1$ to meet the larger circle at X_1. Drop a perpendicular from X_1 onto the boost axis to meet it at D_1. Extend $A_1 B_1$ to meet $X_1 D_1$ at C_1. This point is on the ellipse. This construction may be used to obtain points on the ellipse *without* drawing it. Then $A_1 C_1 = O'D_1 = \gamma_u P'c \cos \theta_1'$.

6. Mark a point O on the major axis of the ellipse that $OO' = \gamma_u \beta_u E'$. This makes $OD_1 = \gamma_u (P_L'c + \beta_u E')$. From the Lorentz transformation (equations (6.17)), this is $P_L c$.

7. Since $O'A_1 = D_1 C_1 = P_T'c = P_T c$ (equation (6.17)), it follows that $OC_1 = \mathbf{P}c$ and the angle $\angle C_1 OO' = \theta_1$, the angle $\mathbf{P}$ makes with the boost axis in Σ.

8. Consider the *same* particle but moving so that it makes an angle with the boost direction $\theta_2' > 90°$ in Σ'. Then $O'B_2 = \mathbf{P}'c$ and $P_L'c < 0$. The step in 5 above, where $A_1 B_1$ was extended to C_1, has to become $A_2 B_2$ extended to C_2 on the ellipse. (Use the same construction.) Then by similar reasoning OC_2 is the momentum (in Σ) after the boost. And the angle $\angle C_2 OO' = \theta_2$ is the angle that momentum makes with the boost axis. And $O'A_2 = D_2 C_2 = P_T c = P_T'c$ for this particle at θ_2' in Σ'.

9. It is important to note that although the diagram connects momenta before and after the boost, the diagrams are not independent of particle mass (unless it is zero) which comes into the construction through the value of E' in $\gamma_u \beta_u E'$.

It is easy to see how this direction and momentum will vary as the angle θ' is changed. Thus this representation shows properties of Lorentz boosts which may be found from the actual equations but are sometimes more striking shown in this way. Here is an example. Consider a particle 'a' with mass, moving with speed u, which emits a photon

$$a \rightarrow b + \gamma$$

Label: 1   2 .

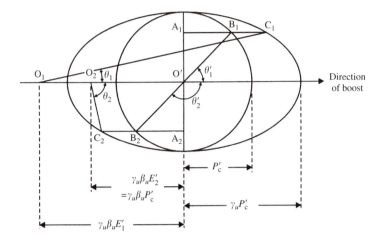

Figure 6.7 The graphical representation of the transformation of the momenta of 'b'
and γ from Σ' to Σ in the decay

$$a \rightarrow b + \gamma,$$

where 'a' is stationary in Σ' and moving with speed u in Σ. The decay
products 'b' and γ have equal and opposite momenta $(P'c)$ in the rest
frame of 'a'. Subscript 2 applies to the photon $(E'_2, \mathbf{P}'c)$ and subscript 1
to the particle 'b' $(E'_1, -\mathbf{P}'c)$. In Σ the vector momentum of the photon
is given by O_2C_2 and of the particle 'b' by O_1C_1.

In Figure 6.7, $O'B_2$ typically represents the photon's momentum vector $\mathbf{P}'c$, and
$O'B_1$ represents the equal but opposite recoil momentum of the particle 'b', both in
the rest frame of particle 'a' (Σ'). The construction follows the procedure described
in the caption to Figure 6.6.

1   We use the bottom half of this figure to demonstrate the transformation for
     the photon which has energy $E' = P'c$. The point O_2 is given by $O_2O' =
     \gamma_u \beta_u E' = \gamma_u \beta_u P'c = \gamma_u \times$ (the semi-major axis). Therefore the point O_2 is
     inside the ellipse. In fact, O_2 is at a focus of the ellipse and this is the case
     for all massless particles. The point C_2 is found by extending the semi-chord
     A_2B_2 to meet the ellipse at C_2. The direction and the momentum (in Σ) of the
     emitted photon is represented by O_2C_2. If $\theta' = 180°$, the photon (in Σ) will
     be at $180°$ to the direction of the boost however close u is to c. Of course, we
     know this has to be the case because the speed of light is always independent
     of the speed of the source.
2   We use the top part of the diagram for particle 'b'. If it has mass, it can happen
     that the point O_1, found setting $O_1O' = \gamma_u \beta_u E'$ (where E' is the energy of 'b'
     in Σ'), is outside the ellipse. In that situation the angle of the momentum of
     'b', O_1C_1, relative to the direction of the boost, is always less than a certain
     maximum, wherever C_1 is on the ellipse. That maximum occurs if O_1C_1 is a
     tangent to the ellipse.

Of course, this graphical method cannot give precise results. However, knowing where the major points are on the axes will show how the final momenta are oriented as θ' is varied. This is something not easily deduced from the transformation equations. Try drawing the graphical construction by answering Problem 6.13.

*6.8 Lorentz boosts at work in particle decays

Figure 3.4 suggests an example of a boost that might be involved in considering the kinematics of the reaction shown. For convenience, we reproduce (Figure 6.8) the sketch of the interesting tracks from that bubble chamber photograph.

In Figure 6.8 the electrically neutral particle (label Λ) moved from vertex V_0 to vertex V_2 where it decayed in flight. The two decay products diverged from V_2:

$$\Lambda \to p + \pi^-.$$

In the laboratory the momenta of the proton and of the pion were measured by the three-dimensional track reconstruction and curvature in a magnetic field. The angular distribution of the proton from the decay of the Λ particle in its rest frame over many events was once of particular research interest. The determination of that distribution might have proceeded as follows. For each event:

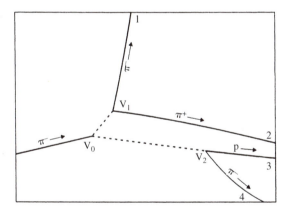

Figure 6.8 A repeat of the labelled line sketch of the particle reaction shown as a bubble chamber photograph in Figure 3.4. The reaction is (at V_0)

$$\pi^- + p \to K^0 + \Lambda,$$

with two decays in flight. At V_2

$$\Lambda \to p + \pi^-$$

and at V_1

$$K^0 \to \pi^+ + \pi^-.$$

a   Calculate the mass of the Λ (equation 3.10):

$$M_\Lambda c^2 = \sqrt{(E_p + E_\pi)^2 - (\mathbf{P}_p + \mathbf{P}_\pi) \cdot (\mathbf{P}_p + \mathbf{P}_\pi)c^2}.$$

b   Calculate the velocity and Lorentz factor of the Λ:

$$\boldsymbol{\beta}_\Lambda = \mathbf{v}_\Lambda/c = (\mathbf{P}_p + \mathbf{P}_\pi)c/(E_p + E_\pi), \quad \gamma_\Lambda = (E_p + E_\pi)/(M_\Lambda c^2).$$

c   Set up the inertial frame Σ in the laboratory with its z-axis along $\mathbf{v}_\Lambda$, and its x-axis in the direction parallel to the vector $\mathbf{P}_\Lambda \times \mathbf{P}_K$ which is normal to the Λ–K production plane. (That plane is defined as containing $\mathbf{P}_\Lambda$, the momentum of the Λ, and $\mathbf{P}_K$ which is the momentum of the particle, a neutral kaon (K^0), that moves from V_0 to V_1.)

d   Make a Lorentz boost of the energies and momenta of proton and of the pion along the z-axis with speed $-v_\Lambda$, that is, into a rest frame (Σ') of the Λ. Use equation (6.5), with $\beta_u = -v_\Lambda/c$. That will yield the momenta of the p ($\mathbf{P}_p$) and of the π^- ($\mathbf{P}_\pi$) in that frame. They should be equal and opposite. The angle between the direction $\mathbf{P}_p$ of the proton and the x'-axis in Σ' can be calculated.

e   Over many events, it is possible to build up an angular distribution of the decay proton direction with respect to the x'-axis. (That direction is normal to the Λ–K^0 production plane in Σ'.)

It was that distribution which demonstrated the non-conservation of parity in the decay of the Λ (see Perkins[2]).

**6.9   Collision kinematics

A laboratory is the place where collisions are organised and their products observed. On the atomic or sub-atomic scale we are limited to organising collisions between two particles. The products of such collisions are not limited and can be two or more in number (Figure 6.5). Even if there are only two they need not be the same as the colliding particles, for example, Figure 3.4. The change from initial to final state in a collision is governed not only by conservation of energy and momentum but also by other conservation laws such as that of electric charge. And, of course, in the change from initial to final state the velocity and energy of the centre-of-mass remain the same.

The centre-of-mass (Section 3.12) of the colliding particles may be the most convenient frame for an analysis of the collision dynamics. Then it is necessary to boost the observed energy–momentum of each of the initial and final state particles from the laboratory to the centre-of-mass. That requires a knowledge of the centre-of-mass velocity, $\mathbf{v}_{cm}$ in the laboratory (equation (3.19)). The transformations are, in principle, no different from those performed in the preceding sections. The choice of coordinate systems may require some care.

2 Perkins, D. H., *Introduction to high energy physics*, 3rd edn, Menlo Park, Addison-Wesley, 1987, pp. 220–2.

In Sections 3.11–3.13 we covered some ground about the centre-of-mass of a group of particles in various circumstances. We now put together what is needed for the construction of boosts in the environment of two-body collisions:

1   In the laboratory frame Σ_{lab} the two particles (masses M_1, M_2) have energy–momentum $(E_1, \mathbf{P}_1)$ and $(E_2, \mathbf{P}_2)$.

2   The centre-of-mass energy W is given by (equation (3.18))

$$W^2 = (E_1 + E_2)^2 - (\mathbf{P}_1 + \mathbf{P}_2) \cdot (\mathbf{P}_1 + \mathbf{P}_2)c^2.$$

3   The energies and momenta (primed quantities) in the frame Σ_{cm} satisfy (equation (3.5)):

$$E_1' = \frac{W^2 + (M_1 c^2)^2 - (M_2 c^2)^2}{2W}, \quad E_2' = \frac{W^2 + (M_2 c^2)^2 - (M_1 c^2)^2}{2W},$$

$$W = E_1' + E_2', \quad \mathbf{P}_1' = -\mathbf{P}_2',$$

$$P_1 = P_2 = \sqrt{(E_1')^2 - (M_1 c^2)^2} = \sqrt{(E_2')^2 - (M_2 c^2)^2}.$$

4   The velocity, $\mathbf{u}_{\text{cm}}$, of the centre-of-mass frame Σ_{cm} in the laboratory frame Σ_{lab} is given by (equation (3.19))

$$\mathbf{u}_{\text{cm}}/c = \boldsymbol{\beta}_u = (\mathbf{P}_1 + \mathbf{P}_2)c/(E_1 + E_2),$$

and the Lorentz factor is given by (equation (3.20))

$$\gamma_u = (E_1 + E_2)/W.$$

To boost from Σ_{cm} to Σ_{lab} use equation (6.4) or (6.5) with the speed parameter β_u. As usual, for the reverse boost use the same equations but with parameter $-\beta_u$, and the centre-of-mass and laboratory kinematic quantities interchanged.

Here are some examples of situations involving two-body collisions:

a   Fixed target collisions were considered specifically in Section 3.12 and equations (3.21)–(3.24) give the necessary formulae for the construction of boosts.

b   Electron–positron colliders with $\mathbf{P}_1 = -\mathbf{P}_2$. Thus the laboratory and the centre-of-mass coincide! (LEP at CERN, 104 on 104 GeV, and others at lower energy.) The Tevatron collider (proton–antiproton collisions at 1000 on 1000 GeV) has the same property (Section 4.13).

c   For some years CERN operated the Intersecting Storage Ring. For research purposes the two stored beams of 31 GeV/c protons intersected, not quite head-on, but at an angle of 14.8°. See Problem 6.15.

d   The HERA collider at the DESY laboratory (Section 4.13) intersects a 30 GeV positron beam with an 820 GeV proton beam at an angle of 0°.

e   The Stanford Linear Accelerator Centre (SLAC) in California and the KEK
Laboratory in Japan are each operating an asymmetric e^+e^- collider. Both
will be tunable to a centre-of-mass energy of 10.58 GeV (see Problem 6.16).
That will lead to the *formation* (Section 3.8) of the vector meson $\Upsilon(10580)$.
(The number in parenthesis is the mass in MeV/c^2.) That meson decays to
give B-mesons that in turn decay to other mesons. By having an asymmetric
primary collision (9 GeV e^- on 3.1 GeV e^+ at SLAC) these B-mesons will
have greater laboratory speeds than would be the case with a symmetric collider. That speed taken with the effect of time dilation (Sections 7.10–7.14)
greatly increases the probability that these particles will move far enough from
the original collision to allow reliable detection of their decay products and
precise measurements of their decay properties (Problem 7.7). Such measurements are one of the primary research objectives for these colliders and the
results will test cosmological theories that implicate the B mesons as a cause
of the excess of matter over anti-matter in the universe.[3]

*6.10   Summary

It is interesting to note that the applications of relativistic kinematics are in the
area of the very large in astrophysics and of the very small in nuclear and particle
physics. There are many aspects of both that we cannot discuss here. Hagedorn,[4]
for example, deals with a range of relativistic problems in the context of particle
physics.

Problems 6.7–6.9 are about the very large. Problems 6.10–6.19 concern the very
small!

Much of relativistic kinematics becomes simpler with the use of four-vectors.
We introduce their application to kinematics in Section 9.2.

Problems

6.1   A Lorentz transformation of the energy and momentum of a particle
from inertial frame Σ' to Σ changes $(E', \mathbf{P}') \rightarrow (E, \mathbf{P})$, as given by
equations (6.3). Show that

$$E^2 - \mathbf{P} \cdot \mathbf{P}c^2 = E'^2 - \mathbf{P}' \cdot \mathbf{P}'c^2.$$

6.2   Show that a there-and-back Lorentz transformation $\Sigma' \rightarrow \Sigma \rightarrow \Sigma'$ of E'
and $\mathbf{P}'$ using equations (6.3), once with u and again with $-u$, restores E'
and $\mathbf{P}'$ to their original values.

3 Quinn, H. R. and Witherell, M. S., 'The asymmetry between matter and antimatter', *Scientific American* **279**(4), 1998, 50–5.
4 Hagedorn, R., *Relativistic kinematics*, New York: W. A. Benjamin, Inc., 1963.

6.3 If

$$
L(\beta_u) = \begin{pmatrix} \gamma_u & 0 & 0 & \gamma_u\beta_u \\ 0 & 1 & 0 & 0 \\ 0 & 0 & 1 & 0 \\ \gamma_u\beta_u & 0 & 0 & \gamma_u \end{pmatrix},
$$

where $\beta_u = u/c$ and $\gamma_u = 1/\sqrt{1-\beta_u^2}$, confirm that

$$
L(\beta_u)\,L(-\beta_u) = I,
$$

where I is the four by four unit matrix.

6.4 There are two particles in a Cartesian inertial frame Σ (Figure 6.1) having energy and momentum E_i, P_i where $i = 1, 2$. The same particles observed from an inertial frame Σ' have energies and momenta E', P'. Show that

$$
E_1 E_2 - P_1 \cdot P_2 c^2 = E_1' E_2' - P_1' \cdot P_2' c^2.
$$

6.5 In the year 2050 a spaceship shuttle operates between Earth and Mars. Each spaceship is equipped with identical monochromatic (wavelength λ) head and tail lights. Their cruising speed v is such that to the controllers on Earth the head lights of a returning shuttle appear green (500 nm) and the tail lights of a Mars-bound shuttle appear red (600 nm).

1 What is the wavelength λ of the headlights?
2 What is the speed v of the shuttles?

6.6 Ions in an accelerator beam have a uniform velocity and emit a spectral line which is observed in two different directions in the laboratory system of coordinates. The wavelength observed perpendicular to the beam is 474.4 nm; observation parallel with the beam direction gives a wavelength 359.5 nm. Find the beam speed and the wavelength of the spectral line in the ion rest frame.

6.7 A binary star system has emitted a brief, massive jet of very hot gas. This gas is a strong source of radio waves and a collaboration of radio telescopes is able to track this jet across the sky for a year. The result is that it appears to be moving at about 2.5 times the speed of light (a super-luminal speed). The radio astronomers suggest that the jet is actual moving at about $0.97c$ and at an angle of $5°$ to the line from the galaxy to the Earth. Show that this explanation is reasonable. (The solution to this problem does not require special relativity!)

6.8   A galaxy with a small speed of recession from the Earth has an active
nucleus. It has emitted two brief jets of hot material with the same speed v,
but in opposite directions. See the figure.

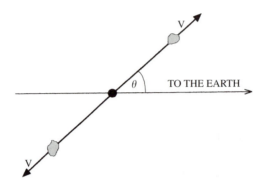

In each jet an atomic species is emitting light of proper wavelength λ_0. The
light from the two jets is observed from the Earth (which may be assumed
to be at a large distance compared with the distance of each jet from the
galactic nucleus). The wavelengths measured are λ_+ for the approaching
source and λ_- for the receding source. Show that

$$\lambda_\pm/\lambda_0 = \gamma_v(1 \mp \{v/c\}\cos\theta),$$

where $\gamma_v = 1/\sqrt{1 - v^2/c^2}$.
   If $\lambda_0 = 448.1$ nm, $\lambda_+ = 420.2$ nm, and $\lambda_- = 700.1$ nm, find v and θ.
   In some cases the receding source is difficult to observe. Suggest a reason
for this fact.

6.9   An absorption line in singly ionised calcium is measured to be at a wavelength of 394 nm measured for stationary absorber. The same absorption
line in the spectrum of the star Hydra occurs at 475 nm. What can be deduced
about the motion of Hydra with respect to the Earth?

6.10   The neutral pion, rest mass M_π, decays into two photons:

$$\pi^0 \to \gamma + \gamma.$$

What is the minimum angle between the direction of the two photons that
may be observed in the laboratory where the neutral pions have a speed v?

6.11   A neutral pion, rest mass M_π, moves along the z axis with speed βc in the
laboratory (Σ) and undergoes decay into two photons. Obtain expressions
for the laboratory energy of a photon if it is

   i   emitted perpendicular to the z-axis in the π^0 rest frame,
   ii observed perpendicular to the z-axis in the laboratory,
   iii emitted along the direction of motion of the π^0.

In each case find the energy and direction of the second photon. Calculate the numerical results for the case of $\beta = 0.9$ ($M_\pi = 134.97\,\text{MeV}/c^2$).

6.12   From equation (6.18) we can define

$$(M_T c^2)^2 \equiv (Mc^2)^2 + (P_T c)^2 = E^2 - (P_L c)^2.$$

Then

$$\frac{E^2}{(M_T c^2)^2} - \frac{(P_L c)^2}{(M_T c^2)^2} = 1,$$

which may be written

$$(\cosh y)^2 - (\sinh y)^2 = 1.$$

Show that

$$y = \frac{1}{2}\ln\left(\frac{E + P_L c}{E - P_L c}\right) = \ln\left(\frac{E + P_L c}{M_T c^2}\right)$$

$$= \text{the rapidity (Section 6.6).}$$

Show that if the rapidity in Σ' is y' then the rapidity in Σ is y given by

$$y = y' + \tanh^{-1}\beta_u.$$

6.13   A decay mode of the neutral kaon is

$$K^0 \quad \rightarrow \pi^+ \quad + \quad \pi^-$$

Masses in MeV/c^2:   497.7   139.6   139.6.

Calculate all the necessary kinematic parameters and draw the graphical representation of the Lorentz boost for the two decay pions from the rest frame of the K^0 to the laboratory where the K^0 has momentum 300 MeV/c. Consider one pion emitted, in the laboratory, at 90° to the direction of the K^0. Use the diagram to estimate the values in the laboratory of

1   the momentum of that pion, and
2   the direction and momentum of the other pion.

6.14   In Section 3.3 we discussed the decay of positive pions:

$$\pi^+ \rightarrow \quad \mu^+ \quad + \quad \nu$$

Masses in MeV/c^2:   139.6   105.7   0.

Consider a beam of positive pions of momentum 200 MeV/c in a laboratory. One of these pions decays, emitting a muon in its rest frame at 90° to the direction in which the laboratory appears to be moving in the rest frame. Find the total energy of the muon in the laboratory and the angle (with respect to the direction of motion of the parent pion) at which it will be observed in the laboratory. Find also

1 the maximum angle at which muons will be observed in the laboratory and their energy at this angle,
2 the least energy muons will have in the laboratory and the angle at which they will be observed.

6.15   The CERN Intersecting Storage Rings operated with two stored beams of 31.44 GeV/c momentum protons. These beams intersected at an angle of 14.8° (165.2° between the directions of the incident momenta). Calculate

1 the energy in the centre-of-mass of the collisions, and
2 the magnitude of the velocity of this centre-of-mass in the laboratory and its direction with respect to the colliding beams.

A low energy charged pion is produced with a momentum of 300 MeV/c in the centre-of-mass. It is moving, in that inertial frame, in a direction perpendicular to the plane containing the incident protons. What is its momentum in the laboratory?

6.16   Here is a problem about an asymmetric collider of the type that is now operating. In the laboratory positrons of energy 3.1 GeV collide head-on with electrons of energy E. What value of E is required to ensure the formation of Υ mesons ($M_\Upsilon c^2 = 10580$ MeV) in some collisions:

$$e^+ + e^- \rightarrow \Upsilon.$$

Calculate the energy and momentum of the Υ mesons in the laboratory. This problem continues in Problem 7.7. (Neglect the mass of the electron and positron.)

6.17   Photons of wavelength 400 nm meet head-on a beam of electrons of energy 10 GeV. Some photons are scattered through 180° in collisions with electrons. Calculate the energy of these scattered photons.

6.18   Consider the three body decay

$$a \rightarrow 1 + 2 + 3,$$

in the rest frame of 'a'. In Section 3.4 we showed the energy of 3 had to be between the limits given by

$$M_3 c^2 \leq E_3 \leq \frac{(M_a c^2)^2 + (M_3 c^2)^2 - (M_1 c^2 + M_2 c^2)^2}{2 M_a c^2}.$$

Assuming E_3 lies in the permitted range what is the range of allowed values of E_2? The reader is invited to show the correctness of formulae in a step-by-step guide.

Step 1. Consider the system of particles 1 and 2. The energy in the centre-of-mass can be thought of as the energy associated with the mass of the system. Show that the (rest) mass M_{12} of the system $1 + 2$ recoiling from 3 is given by

$$(M_{12}c^2)^2 = (M_a c^2)^2 + (M_3 c^2)^2 - 2E_3 M_a c^2.$$

Step 2. Show that the energy and momentum of the $1+2$ system in the rest frame of 'a' are given by

$$E_{12} = M_a c^2 - E_3 \quad \text{and} \quad P_{12}c = \sqrt{E_{12}^2 - (M_{12}c^2)^2}.$$

Step 3. Show that the rest frame (= centre-of-mass) of the $1 + 2$ system has a speed v and Lorentz factor γ in the rest frame of 'a' that are given by

$$v/c = \beta = P_{12}c/E_{12} \quad \text{and} \quad \gamma = E_{12}/M_{12}c^2.$$

Step 4. Show that the energy E^* and momentum P^* of particle 2 in the rest frame of the $1 + 2$ system are

$$E^* = \frac{(M_{12}c^2)^2 + (M_2 c^2)^2 - (M_1 c^2)^2}{2M_{12}c^2}, \quad P^*c = \sqrt{E^{*2} - (M_2 c^2)^2}.$$

Step 5. In the rest frame of 'a' the reverse of the direction of the momentum of particle 3 is the direction of the velocity of the $1 + 2$ system. Suppose particle 2 has its momentum at an angle θ^* (in the $1 + 2$ rest frame) to this direction. Show that the energy E_2 and momentum P_2 of 2 in the rest frame of 'a' are given by

$$E_2 = \gamma(E^* + \beta P^* c \cos \theta^*),$$
$$P_2 c \cos \theta = \gamma(P^* c \cos \theta^* + \beta E^*),$$
$$P_2 c \sin \theta = P^* c \sin \theta^*,$$

where θ is the angle $\mathbf{P}_2$ makes with the reverse of the direction of $\mathbf{P}_3$ in the rest frame of 'a'.

Step 6. Show that the allowed values of E_2 are in the range given by

$$\gamma(E^* - \beta P^* c) \leq E_2 \leq \gamma(E^* + \beta P^* c).$$

Step 7. Remember these formula carry a hidden E_3 that was chosen to be inside its allowed range of values.

Step 8. Given any allowed E_2 and E_3, then E_1 and $\mathbf{P}_1$ in the rest frame of 'a' may be found from the conservation of energy and momentum.

6.19   Consider the decay

$$K^+ \quad \rightarrow \quad \pi^+ \quad + \quad \pi^+ \quad + \quad \pi^-$$

Masses in MeV/c^2:   493.68   139.57

Suppose the kinetic energy of the third pion is 20 MeV. Using the results of Problem 6.18, find the greatest and least kinematically allowed values of the kinetic energy of one of the other pions in the rest frame of the K^+.

7 The Lorentz transformation of time–space coordinates

**7.1 Introduction

In this chapter we introduce the Lorentz transformations for the time and space coordinates of an event. Our presentation will be analogous to the presentation of the transformation of energy and momentum that we made in Section 6.2.

We shall use our standard arrangement of reference frames. But a word of warning! Unmoving diagrams on a piece of paper must be interpreted with care. The existence of point-like events marked on the paper cannot imply that there is a common or universal time connecting them.

It is now that the results of special relativity become more challenging. We trust that the reader has found the results of the Lorentz transformation of energy and momentum of a particle to be reasonable. The matrix that transforms the time and space coordinates of an event is identical to that (**L**, Problem 6.3 and equation (6.5)) which transforms energy and momentum. However, the consequences for time–space can be counter intuitive.

**7.2 The Lorentz transformation of time–space coordinates

We return to our standard arrangement which we reproduce in Figure 7.1. Events and coordinates have various representations. Here are some guidelines to our notation:

a   Events are labelled $= \mathbb{O}, \mathscr{P}, \mathscr{Q}, \mathscr{R}, \ldots$.
b   In the inertial frame Σ these events have coordinates $(t, \mathbf{r})_\mathbb{O}, (t, \mathbf{r})_\mathscr{P}, \ldots$, and so on for the script letters used.
c   In the frame Σ' the same events have coordinates $(t', \mathbf{r}')_\mathbb{O}, (t', \mathbf{r}')_\mathscr{P}, \ldots$.
d   The position vector $\mathbf{r}$ has components x, y, z.
e   There is one special event that we shall sometimes use: it is defined by the coincidence of the origins O and O′ and sets the zero of time in Σ and in Σ'. If given the label $\mathbb{O}$, it has, therefore, coordinates $(t, \mathbf{r})_\mathbb{O} = (0, 0) = (t', \mathbf{r}')_\mathbb{O}$.
f   Sometimes we may spell out the coordinates x, y, z of $\mathbf{r}$, or drop all but one.
g   If necessary coordinates may be displayed in simple tables in order to make an argument easier to follow.

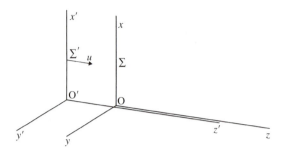

Figure 7.1 Our standard arrangement (Figure 5.1) of two inertial reference frames reproduced here for convenience. Recall that Σ' is moving with speed u in Σ.

h   Mathematically the Lorentz transformation of coordinates is most clearly stated in a matrix equation. Then the coordinates are represented by a column matrix.

i   In Section 7.4 we shall begin routinely to use ct in place of t in order to give all coordinates the same dimension.

   As before u is the speed of Σ' in Σ. Given the event $\mathbb{O}$, the Galilean transformation (equations (5.2) and (5.3)) of the coordinates of any other event are

$$t = t',$$
$$x = x',$$
$$y = y',$$
$$z = z' + ut'.$$
(7.1)

We now write down the Lorentz transformations:

$$t = \gamma_u (t' + \{uz'\}/c^2),$$
$$x = x',$$
$$y = y',$$
$$z = \gamma_u (z' + ut'),$$
(7.2)

where, as before in Section 6.2,

$$\gamma_u = 1/\sqrt{1 - u^2/c^2}.$$

These equations are very similar to those for the energy and momentum transformations (equations (6.3)) and we look for some analogies in their properties:

1   The invariant in these transformations is

$$(ct)^2 - \mathbf{r} \cdot \mathbf{r} = (ct')^2 - \mathbf{r}' \cdot \mathbf{r}'.$$
(7.3)

The reader should check this equality (Problem 7.1(a)).

2   The inverse transformation that takes coordinates in Σ to those in Σ', may be found by interchanging $(t, \mathbf{r}) \leftrightarrow (t', \mathbf{r}')$ and changing $u \rightarrow -u$ in equation (7.2). A 'there-and-back' transformation returns the coordinates to their original value (Problem 7.1(b)).

3   The non-relativistic limit is easily found by taking $c \rightarrow \infty$. Then $\gamma_u \rightarrow 1$, $uz'/c^2 \rightarrow 0$ and the Galilean equation (7.1) are regained.

4   Consider the origin O'. It has $\mathbf{r}' = 0$ for all t'. Substituting in equation (8.2) we find that the coordinates in Σ are

$$t = \gamma_u t',$$

$$x = 0,$$

$$y = 0,$$

$$z = \gamma_u u t'.$$

Eliminating t' we find

$$z = ut,$$

which is just what we expect for the position of a point, the origin O' of the coordinates in Σ', moving with speed u along the z axis in Σ.

However, the Lorentz transformations of the event coordinates are more fundamental than those of the energy–momentum of a particle. Logically they precede the latter. We shall give a derivation and a discussion of both transformations in Chapter 8. Now we must check that the former satisfy the postulates of special relativity (Section 5.5).

**7.3   Conformity with the postulates

The second postulate of special relativity states that the speed of light in empty space is always the same. Suppose event $\mathcal{O}$ (see the previous section) is the event in which a flash of light is emitted at a time $t' = 0$ from a source stationary at the origin O' in Σ'. Let the event $\mathcal{P}$ be that in which the flash is detected at a point a distance $\mathbf{r}'$ from O' at a time t'. The observer in Σ' deduces the speed of light $c = r'/t'$ and expresses this as

$$(ct')^2 - \mathbf{r}' \cdot \mathbf{r}' = 0.$$

Note that for this Σ' observer the light source is stationary. An observer in Σ using identical clocks and distance measuring devices as available in Σ', on the same events, finds that (equation (7.3))

$$(ct)^2 - \mathbf{r} \cdot \mathbf{r} = 0,$$

or, taking the magnitudes

$$r/t = c.$$

Now r/t will be the speed of light found by the observer in Σ who records the coordinates of events $\mathcal{O}$ and $\mathcal{P}$. It is the same as that found in Σ', independent of the direction of $\mathbf{r}'$, or of the speed of Σ' in Σ. Thus the Σ observer can state that the speed of light emitted by a source is independent of velocity of that source. The second postulate is satisfied by the Lorentz transformations.

This is, of course, as it must be: Einstein derived the Lorentz transformations by requiring that they satisfy this postulate.

The first postulate requires that the laws of physics are the same in all inertial frames. In deriving the consequences of the postulates, Einstein required that conservation of energy and Newton's laws of motion apply in all frames. (He restricted the application of the second to particles instantaneously at rest.) His second postulate also allowed him to require that Maxwell's equations were covariant (Section 5.8) and from this he was able to derive the transformation equations for the electric and magnetic fields. From that followed the relativistic kinematics and electrodynamics that we have discussed in Chapters 2–4, with results that have successfully passed all experimental tests. In fact, Lorentz and others knew that Maxwell's equations were covariant under these transformations. Einstein's success was to find the Lorentz transformations from his two postulates but without appeal to the need of covariance for Maxwell's equations.

We shall examine how Maxwell's equations (Section 9.9) and the laws of conservation of energy and momentum (Section 8.12) do satisfy the first postulate.

**7.4   A simplification in notation

The energy–momentum transformations took a simple form if we always associate a c with $\mathbf{P}$. This made the units of the transformed quantities $E, \mathbf{P}c$ identical, namely energy. We now do the same for time–space. We link a c with t so that the units of the transformed coordinates $(ct, \mathbf{r})$ are the same, namely distance. Many books use units in which $c = 1$. We do not do this but remain consistent with the policy set out in Section 3.1. We can now rewrite the Lorentz transformation equation (7.2) as:

$$\begin{aligned}
t &= \gamma_u(t' + \{uz'\}/c^2) &\rightarrow\quad ct &= \gamma_u(ct' + \beta_u z'), \\
x &= x' &\rightarrow\quad x &= x', \\
y &= y' &\rightarrow\quad y &= y', \\
z &= \gamma_u(z' + ut') &\rightarrow\quad z &= \gamma_u(z' + \beta_u ct'),
\end{aligned} \tag{7.4}$$

where, as before

$$\beta_u = u/c \quad \text{and} \quad \gamma_u = 1/\sqrt{1 - \beta_u^2}.$$

Or, in matrix form

$$
\begin{pmatrix} ct \\ x \\ y \\ z \end{pmatrix} = \begin{pmatrix} \gamma_u & 0 & 0 & \gamma_u \beta_u \\ 0 & 1 & 0 & 0 \\ 0 & 0 & 1 & 0 \\ \gamma_u \beta_u & 0 & 1 & \gamma_u \end{pmatrix} \begin{pmatrix} ct' \\ x' \\ y' \\ z' \end{pmatrix}.
$$

(7.5)

The square matrix in this equation is identical to that, $L(\beta_u)$, in equation (6.5). Thus the time–space coordinates ct', x', y', z' are transformed as are the energy momentum E', $P'_x c$, $P'_y c$, $P'_z c$. We shall prove this in Section 8.10. It is an important simplification as it allows easy recall of a matrix frequently needed and because it points to an emerging unity in the structure of special relativity. We shall explore this unity in Chapters 8 and 9. And, of course, the matrix that transforms $(ct, \mathbf{r})$ in Σ to $(ct', \mathbf{r}')$ in Σ' is just $L(-\beta_u)$, as in Section 6.2.

*7.5 A world line

We can imagine a particle moving through space. Its space coordinates will depend on the time and therefore its motion may be represented by a continuous line in a four dimensional space ct, x, y, z. This is called a **world line**. We can define a line element dS given by

$$
dS = +\sqrt{(dx)^2 + (dy)^2 + (dz)^2}.
$$

Thus dS is the magnitude of the distance between two points for which the coordinate separations are dx, dy and dz. Then there are some restrictions on dS/dt for world lines:

1 Since the particle cannot go backwards in time its world line must have

$$
\frac{dS}{dt} \geq 0.
$$

2 If the particle has rest mass it cannot move with a speed equal to or greater than the speed of light therefore

$$
\frac{dS}{dt} < c.
$$

3 A photon moves with the speed of light so a photon world line must have

$$
\frac{dS}{dt} = c.
$$

We cannot represent world lines in four dimensions on paper. Figure 7.2 shows two world lines in the two dimensions ct, z. The line at $45°$ to both axes is a world line for a photon moving along the z-axis. The other would be the world line for a particle moving along the z-axis with a non-uniform speed.

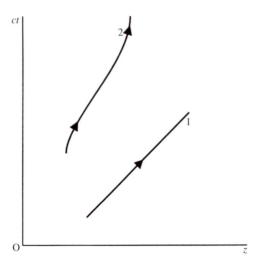

Figure 7.2 World lines in two dimensions *ct* and *z*. That labelled 1 is that for a photon travelling along the *z*-axis. That labelled 2 is for a particle with rest mass moving with a non-uniform speed along the *z*-axis.

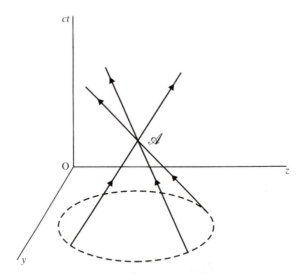

Figure 7.3 Photon world lines passing through event *A* in three dimensions *ct*, *y*, and *z*. We have tried to draw them as they would appear if each was on a cone of half angle 45°, with its axis parallel to the *ct*-axis.

We can attempt to represent world lines in three of four dimensions. In Figure 7.3 we suppress the *x*-axis of a Cartesian three-dimensional spatial reference frame and replace it with the *ct*-axis. An event *A* occurs. There is a family of photon world lines that pass through *A* each at 45° to the *ct*-axis, and to the *yz* plane.

They generate a cone with its apex at $\mathscr{A}$. This is the **light cone** at $\mathscr{A}$. It has two branches: the first extends into the future of $\mathscr{A}$ ($ct > ct_{\mathscr{A}}$) and the second branch converges from the past of $\mathscr{A}$ ($ct < ct_{\mathscr{A}}$). In reality a photon represented by one of these world lines is moving in the yz-plane, and is present at event $\mathscr{A}$.

*7.6   The interval

The Lorentz transformations are linear in the coordinates. Therefore, if two events $\mathscr{P}$ and $\mathscr{Q}$ have coordinates $(ct, \mathbf{r})_{\mathscr{P}}$ and $(ct + \Delta ct, \mathbf{r} + \Delta \mathbf{r})_{\mathscr{Q}}$ respectively, then the interval Δct, $\Delta \mathbf{r}$ transforms as do the coordinates themselves. The squared interval $(\Delta s)^2$ is defined by

$$(\Delta s)^2 \equiv (\Delta ct)^2 - (\Delta \mathbf{r} \cdot \Delta \mathbf{r}) \equiv (\Delta ct)^2 - (\Delta x)^2 - (\Delta y)^2 - (\Delta z)^2.$$

Therefore, just as $(ct)^2 - x^2 - y^2 - z^2$ is an invariant, so is $(\Delta s)^2$. There is a convention that the brackets are normally omitted: so

$$\Delta s^2 \equiv (\Delta s)^2 (\neq \Delta (s)^2).$$

and so we write

$$\Delta s^2 \equiv \Delta ct^2 - \Delta x^2 - \Delta y^2 - \Delta z^2. \tag{7.6}$$

(In the same spirit, there is an invariant, infinitesimal squared interval defined by

$$ds^2 \equiv dct^2 - dx^2 - dy^2 - dz^2.$$

The infinitesimal intervals found for curved space–time contain information about the curvature and therefore play an important role in General Relativity. The curvature associated with any interval in an inertial frame is zero.)

The value of Δs^2 allows an important classification of the relation between the events $\mathscr{P}$ and $\mathscr{Q}$ when they are separated by the interval Δct, $\Delta \mathbf{r}$:

1   $\Delta s^2 = 0$ and $\Delta t > 0$: $\mathscr{Q}$ lies on the future light cone of $\mathscr{P}$.
2   $\Delta s^2 = 0$ and $\Delta t < 0$: $\mathscr{Q}$ lies on the past light cone of $\mathscr{P}$.
3   $\Delta s^2 > 0$ and $\Delta t > 0$: $\mathscr{Q}$ lies inside the future light cone of $\mathscr{P}$.
4   $\Delta s^2 > 0$ and $\Delta t < 0$: $\mathscr{Q}$ lies inside the past light cone of $\mathscr{P}$.
5   $\Delta s^2 < 0$: $\mathscr{Q}$ lies outside the light cone of $\mathscr{P}$.

In the cases 3 and 4 which have $\Delta s^2 > 0$, $\mathscr{Q}$ and $\mathscr{P}$ are said to be connected by a **time-like interval**. In case 3, $\Delta t > 0$, a light flash from the event $\mathscr{P}$ can reach the space coordinates of $\mathscr{Q}$ before that event occurs. And, of course, the other way around, if the relative timing is reversed. Thus, depending on the time ordering of the events, signals from one event can causally affect the other event, in these cases.

In case 5, the events are said to be connected by a **space-like interval**. No light signal from one event can reach the space position of the other event before it occurs.

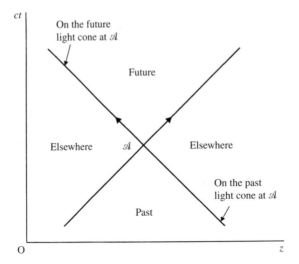

Figure 7.4 This is Figure 7.3 projected onto the *ct–z* plane. All world lines have been removed except those for two photons, both passing through event $\mathcal{A}$, one travelling in the $+z$ direction the other travelling in the $-z$ direction.

In Figure 7.4 we have labelled several regions. With respect to one event, $\mathcal{A}$, a second event is either

a   on the light cone, future (1) or past (2), or
b   in the future (3) or in the past (4), or
c   elsewhere (5).

We can make two statements:

1   Δs^2 is an invariant,
2   if $\Delta s^2 \geq 0$, the time ordering of any two events is the same in all inertial frames (see Problem 7.3).

From these we can conclude that any statement of the kind a, b, or c made about the relative status of any two events is true in all inertial frames.

Consider again event $\mathcal{P}$ having an effect on event $\mathcal{Q}$. The principle of causality leads us to require that event $\mathcal{P}$ occurs before event $\mathcal{Q}$, and relativity requires this to be true in all inertial frames. But such time-ordering is preserved between $\mathcal{P}$ and $\mathcal{Q}$ only if $\mathcal{Q}$ is on or inside the future light cone at $\mathcal{P}$. The time-ordering is not invariant if $\mathcal{P}$ and $\mathcal{Q}$ have a space-like interval. In that situation suppose $\mathcal{P}$ did affect $\mathcal{Q}$ in one inertial frame then in another frame the effect might precede the cause. That is impossible if we stay with causality and relativity. Now the fastest signal that can have an effect on or inside the future light cone is light itself. These facts are taken to mean that no information may be transmitted faster than the speed of light.

**7.7 The absence of simultaneity in special relativity

We return to the loss of simultaneity that was discussed in Section 2.5. We can now do a proper relativistic analysis of the events that involved Einstein's train, Figure 2.2. For convenience we duplicate this picture in Figure 7.5. We assume that the essentials of this situation may be found by treating it as a two dimensional problem with all events lying along the $z(z')$-axis. Therefore we use the standard coordinates but we have separated them to allow less overlap of vital information: see Figure 7.6.

We have the inspector stationary in the inertial frame Σ. The train and the guard are stationary in Σ' which is moving with speed u in Σ. We have the usual orientation of coordinates and the event $\mathbb{O}$ (guard opposite inspector) has coordinates $(ct, \mathbf{r})_{\mathbb{O}} = (ct', \mathbf{r}')_{\mathbb{O}} = (0, 0)$. We organise (!) events $\mathcal{P}$ and $\mathcal{Q}$ (lightning strikes) to take place at $t = 0$ in Σ but at positions $z = +L$ and $z = -L$ respectively. This is simultaneous in Σ in the sense of Einstein's synchronised clocks. The inspector will see these two flashes at a time after $t = 0$, given by the time for the light to travel from $\mathcal{P}$ and from $\mathcal{Q}$ to his position at $z = 0$. Those times are both L/c and the inspector sees the flashes as simultaneous.

Now we transform all event coordinates into the frame Σ'. The results are shown in Figure 7.6. Events $\mathcal{P}$ and $\mathcal{Q}$ are at $z' = +\gamma_u L$ and at $z' = -\gamma_u L$ respectively (recall that the guard is at $z' = 0$). But by clocks synchronised in Σ' the times of these two events are $+\gamma_u \beta_u L/c$ and $-\gamma_u \beta_u L/c$. They are not simultaneous! This completes the formal demonstration of the loss of distant simultaneity but we can complete Einstein's example. In Σ' the guard is equidistant from $\mathcal{P}$ and $\mathcal{Q}$: these non-simultaneous light flashes can only arrive at the guard at different times. That time difference is $2\gamma_u \beta_u L/c$ with the light from $\mathcal{P}$ arriving before that from $\mathcal{Q}$, as anticipated. We hope Figure 7.6 makes the formalities of the situation clear.

Even in the unlikely event of a double lightning strike on a train we can calculate that this effect could never be observed by unaided humans. If $u = 300 \, \text{km h}^{-1}$ (French Train à Grande Vitesse) and $L = 100 \, \text{m}$ the time separation for the guard is only about 2×10^{-13} s.

Figure 7.5 Einstein's train reproduced here from Chapter 5 for convenience. The train guard is leaning out of a window and we catch the scene just as the moving train carries the guard past the inspector standing beside the railway track. Lightning strikes the train at two points simultaneously as observed by the inspector. What does the guard see?

Inertial frame Σ

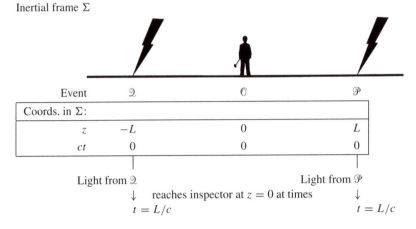

Event	$\mathcal{Q}$	$\mathcal{O}$	$\mathcal{P}$
Coords. in Σ:			
z	$-L$	0	L
ct	0	0	0

Light from $\mathcal{Q}$ Light from $\mathcal{P}$
$\downarrow$ reaches inspector at $z = 0$ at times $\downarrow$
$t = L/c$ $t = L/c$

Inspector sees events $\mathcal{P}$ and $\mathcal{Q}$ as occuring simultaneously.

Lorentz transform event coordinates from Σ to Σ':

$$z' = \gamma_u(z - \beta_u ct), \quad ct' = \gamma_u(ct - \beta_u z),$$

where

$$\gamma_u = \frac{1}{\sqrt{1 - \beta_u^2}}, \quad \beta_u = \frac{u}{c}.$$

Inertial frame Σ'

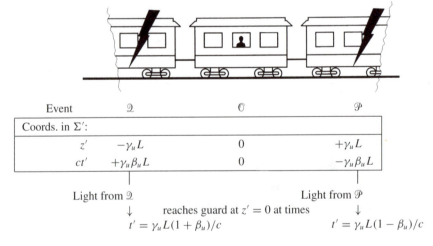

Event	$\mathcal{Q}$	$\mathcal{O}$	$\mathcal{P}$
Coords. in Σ':			
z'	$-\gamma_u L$	0	$+\gamma_u L$
ct'	$+\gamma_u \beta_u L$	0	$-\gamma_u \beta_u L$

Light from $\mathcal{Q}$ Light from $\mathcal{P}$
$\downarrow$ reaches guard at $z' = 0$ at times $\downarrow$
$t' = \gamma_u L(1 + \beta_u)/c$ $t' = \gamma_u L(1 - \beta_u)/c$

Guard sees light from $\mathcal{P}$ a time $2\gamma_u \beta_u L/c$ before light from $\mathcal{Q}$.

Figure 7.6 The train, the inspector, and the guard according to Einstein's special relativity.

We have set up this problem starting with simultaneous events in one frame and calculated the lack of simultaneity observed in a moving frame. We could just as well have done the reverse: set up simultaneous events in the moving frame and calculated the resulting loss of simultaneity in a stationary frame.

The primary conclusion for us is to beware of the lack of simultaneity in relativity. Remember that in an inertial frame there are synchronised clocks at each point. An event has the time coordinate of its local clock. We are once more made aware that an event will not, in general, have the same time coordinate in different inertial frames.

**7.8 The Lorentz–FitzGerald contraction

The loss of simultaneity causes difficulties measuring the dimensions of moving bodies. The question is 'How and when to make measurements?' Consider a simple rod length of length λ at rest in frame Σ'. The 'at rest length' of the rod is called its **proper length**. We place it so that it is coincident with the z'-axis, one end at the origin, the other at $z' = \lambda$ of our standard coordinates. The observer in Σ' consults all his measuring devices and finds the z' coordinates of each end of the rod. This is easy as the rod is stationary and the time of each measurement has no effect. However, this observer is punctilious and knows that all events must be labelled and that the record of every event's coordinates must have four parts (ct', x', y', z'). For want of anything more profound, the two events, $\mathcal{O}$ and $\mathcal{P}$, are given the same time, zero for both. And the record then reads:

$$(ct', x', y', z')_{\mathcal{O}} = (0, 0, 0, 0),$$
$$(ct', x', y', z')_{\mathcal{P}} = (0, 0, 0, +\lambda).$$

Now the frame Σ' is moving with speed u in Σ in the standard way. Therefore to the observer in Σ the rod is lying along the z-axis and moving in the $+z$ direction with speed u. How will this observer measure the length and what will be the result?

The Σ observer cannot find the z position of each end without deciding the time at which measurements should be made. There is only one unambiguous way to proceed. It is to measure the z coordinates of the two ends separately at a time instant that, by arrangement, is the same on two synchronised clocks, one each at an end of the rod at the agreed instant. The observer deduces that the length of the rod is the difference between the z coordinates of the two clocks.

To analyse this situation we make some simplifications of a trivial kind. For our convenience the Σ observer decides that the two measurements shall be made at $t = 0$. When $t = 0 = t'$ the origins of our standard coordinates coincide and the trailing end of the rod is at the origin in Σ. Thus event $\mathcal{O}$ is recorded not only by the Σ' observer but also by the Σ observer. However, the Σ observer's second measurement, that of the spatial position of the leading end of rod at $t = 0$ cannot be of event $\mathcal{P}$ because the simultaneity of events $\mathcal{O}$ and $\mathcal{P}$ in Σ' cannot be reproduced in Σ.

We label this third event 2. We have to find its z coordinate in Σ. We give this coordinate value the symbol L and that is the length that will be measured in Σ. Thus in Σ $(ct, x, y, z)_2 = (0, 0, 0, L)$.

Figure 7.7 will assist us in making the formal transformations. The result is that

$$L = \lambda/\gamma_u. \tag{7.7}$$

Thus any rod length measured in any frame in which the rod is moving lengthwise with speed u is reduced by a factor $\gamma_u^{-1} = \sqrt{1 - u^2/c^2}$ from its proper

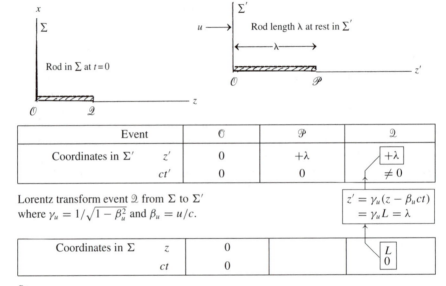

Event	0	$\mathcal{P}$	2
Coordinates in Σ' z'	0	$+\lambda$	$+\lambda$
ct'	0	0	$\neq 0$

Lorentz transform event 2 from Σ to Σ' where $\gamma_u = 1/\sqrt{1 - \beta_u^2}$ and $\beta_u = u/c$.

$z' = \gamma_u(z - \beta_u ct)$
$= \gamma_u L = \lambda$

Coordinates in Σ z	0		L
ct	0		0

Steps:

- The rod is stationary in Σ' and extends from $z' = 0$ to $z' = \lambda$ for all time. The two events 0 and $\mathcal{P}$ have coordinates in Σ' given by $(ct', z')_0 = (0, 0)$ and $(ct', z')_{\mathcal{P}} = (0, \lambda)$ respectively.
- In Σ, event 0 is by definition at the trailing end of the rod and event 2 at the leading end of the rod (which is moving in Σ).
- The events 0 and 2 have coordinates in Σ given by $(ct, z)_0 = (0, 0)$ and $(ct, z)_2 = (0, L)$ where L is to be found.
- Transform coordinates of event 2 in Σ to the $z'(= \lambda)$ coordinate of 2 in Σ' using $z' = \gamma_u(z - \beta_u ct)$ with $\gamma_u = 1/\sqrt{1 - \beta_u^2}$, $\beta_u = u/c$. Note that event 2 is not the same as event $\mathcal{P}$, although both have $z' = \lambda$.
- Result: $\lambda = \gamma_u L$.
- Since $\gamma_u > 1$, the length L observed in Σ is less than the proper length λ observed in Σ': $L = \lambda/\gamma_u$.

Figure 7.7 The details of the transformations that illustrate the Lorentz–FitzGerald contraction.

(= at rest) length. This decrease in apparent length is a consequence of (1) the loss of simultaneity and (2) the prescription for measuring the lengths of moving bodies.

This contraction is called the Lorentz–FitzGerald contraction. Before Einstein, the failure of the Michelson–Morley experiment (Section 10.3) to see the expected effects of the ether caused FitzGerald to suggest that there was a real contraction of a body along the direction of motion through the ether.

There are a couple of loose ends to make tidy.

1   Einstein's result has been derived for a rod of proper length λ. What about a solid body that has width and height as well as length? In our standard coordinates the Lorentz boost along the z-axis has $x = x'$ and $y = y'$. Thus dimensions transverse to the direction of motion are the same to the observer in the frame in which the body is moving as to the observer in the body's rest frame.

2   The use of a rod of proper length λ to derive the Lorentz–FitzGerald contraction is a distraction. It might make readers wonder what is happening in the rod. Consider two permanent markers, stationary on the z'-axis in Σ', a distance λ apart. Observed in Σ these markers are moving with speed u along the z-axis; at identical, synchronised times in Σ the markers will be a distance L/γ_u apart. We do not even need markers! The distance between $z' = 2.57\,\mathrm{m}$ and $z' = 3.94\,\mathrm{m}$ in Σ' will be observed to be $(1.37)/\gamma_u\,\mathrm{m}$ in Σ.

There is a quick way to obtain this 'contraction' result using partial differentiation. The partial derivative

$$\frac{\partial z'}{\partial z}\bigg|_{t}$$

is the ratio of an infinitesimal distance between stationary points on the z'-axis in Σ' to the distance between those points measured on the z-axis at constant, that is to say the same, time in Σ. Now $z' = \gamma_u(z - \beta_u ct)$ from equation (7.4) inverted for the transformation from Σ to Σ' so that

$$\frac{\partial z'}{\partial z}\bigg|_{t} = \gamma_u. \tag{7.8}$$

This gives the same connection between infinitesimal distances in Σ' and Σ as the result in Figure 7.7 gives between λ in Σ' and L in Σ, $\lambda/L = \gamma_u$, which demonstrated the contraction.

But what does happen to a rod? In Σ' the rod has its proper length. There are no new forces that come into play in the frame Σ in which the rod is moving. The theory of the forces responsible for the structure and cohesion of the rod must be Lorentz covariant. If employed in the frame Σ' with independent time variable t', the theory would explain the structure of the rod. Employed in Σ with time variable t, that theory can be applied to the now moving rod. It would show that, at the same time t, all parts of the rod would be closer together longitudinally (= the direction of motion) than the same parts at the time t' in Σ'. Bell[1] has shown this is true for a simple classical model of a hydrogen atom containing an electron moving in an orbit around a proton under the effect of their Coulomb attraction.

7.9   A diversion

We have used 'observer' in the sense of the definition in Section 5.2. This observer organises the measurement of the coordinates throughout their own inertial frame. The time coordinate is given by the clock local to the event; that clock synchronised with all other clocks in the same frame (Section 5.7). What about photographers with colour film in their cameras standing away from the events that are happening? An instantaneous photo does not record events as they are at the same time. The speed of light means that objects distant from the camera are recorded as they were earlier than is the case for objects closer to the camera. Consider a cube, painted with green, self-luminous paint, standing squarely on the origin of coordinates in Σ' (Figure 7.8(a)). A photograph is taken from a point some distance along the y'-axis. Now do the same in Σ so as to obtain a photograph of the cube as its centre passes the origin of coordinates in Σ at the speed u of Σ' in Σ. This to be done at the same distance along what is now the y-axis. Neglecting the real possibility of a collision between the two photographers and without going into details, how do the photographs compare? The first is simple: it will show the nearest face ABCD, in green. The second will show a slant view of the trailing face, and the side ABCD facing the camera shortened in the direction of travel. The effect is as if the cube has been rotated around the x-axis so that two faces become visible to the camera in Σ (Figure 7.8(b)). In addition, the colour will be shifted towards the red part of the visible spectrum. Nearly all this is another story that we cannot tell here.[2] The exception is the red shift; that is the result of the transverse Doppler effect which we met in Section 6.4.

**7.10   Time dilation

For the beginner this is one of the most disturbing aspects of special relativity and is one that sustained controversy over many years. Suppose two events occur. Two separate observations are made from different inertial frames of the time interval between these events. If the two inertial frames have a relative velocity, then the time intervals found are, in general, different. And, if different, the clocks measuring the shorter interval will therefore appear to be running slow with respect to the clocks measuring the longer interval. The loss of simultaneity which we

---

1 Bell, J. S., *Speakable and unspeakable in quantum mechanics*, Cambridge: Cambridge University Press, 1987, pp. 67–80.
2 The interested reader could start with Scott, G. D. and Van Driel, H. J. *American Journal of Physics* **38**, 1970, pp. 971–7, and follow the references therein.

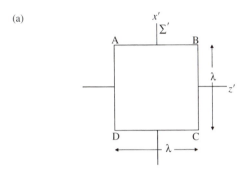

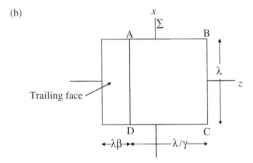

Figure 7.8 A cube is photographed from a point along an axis perpendicular to the face ABCD at a distance such that the solid angle subtended by the cube at the camera is small. (a) shows the appearance of the cube when it is stationary and standing square at the origin of Σ′ and the photograph is taken from the distant point on the y′-axis. In (b) Σ′ has speed u along the z-axis in Σ so that the cube is moving with that speed in Σ. The camera is now on the y-axis at the same distance from the origin as in (a). The photograph is taken so as to record a picture of the cube when its centre is at the origin of Σ. The appearance of the shape of the moving cube is as if it were stationary but rotated through an angle $\sin^{-1}(u/c)$ around the x-axis. On the figure λ is the length of an edge of the cube, $\beta = u/c$, and $\gamma = 1/\sqrt{1 - \beta^2}$.

confirmed in Section 7.7 already implies this result. We now wish to look at this matter in some detail.

We consider a clock stationary at the origin of an inertial frame. With respect to a second inertial frame the first is moving in a manner that causes this clock to move along the z-axis of the second frame. As it does so, it passes a series of synchronised clocks stationary on this z-axis. During the passage of time T' on the moving clock it moves from one clock to another clock among these stationed on the z-axis. These two record the passage of a time T. We need to find how T is related to T'. And do not forget that all clocks are identical (Section 5.2).

Let us deal with this situation formally. We have the usual arrangement of coordinates (Figure 7.1). However everything is happening on the z- and the

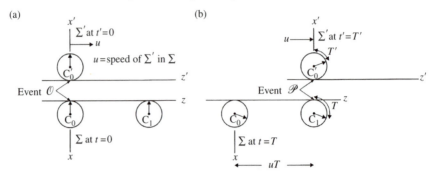

Figure 7.9 The details of the events and transformations demonstrating time dilation. The circles represent clock faces and a single hand in each case indicates a time at an event on the z-axis. Three clocks are involved: one (C_0') is stationary at the origin O' of the coordinates of Σ'. Another (C_0) is stationary at the origin O of Σ and the last (C_1) is stationary at a point on the z-axis of Σ. All these clocks are identical in their operation. All events happen on the z-axis but in the diagram the z- and z'-axes have been separated and the clocks placed so as to make visible their timing at events $\mathcal{O}$ and $\mathcal{P}$.

1 As usual Σ' moves with speed u in Σ and therefore C_0' moves with this speed along the z-axis in Σ.

2 In diagram (a), an event $\mathcal{O}$ has O and O' coincident and therefore C_0 coincident with C_0', and we have both clocks read zero time. Thus $\mathcal{O}$ has the coordinates $(ct, z)_{\mathcal{O}} = (0, 0)$ and $(ct', z')_{\mathcal{O}} = (0, 0)$.

3 Suppose event $\mathcal{P}$ occurs when C_0' reads time T' and reaches clock C_1 on the z-axis in Σ as in (b). Let C_1 read T at event $\mathcal{P}$ and therefore it is at a distance uT from O. Thus $(ct, z)_{\mathcal{P}} = (cT, uT)$ and $(ct', z')_{\mathcal{P}} = (cT', 0)$. All these event coordinates are summarised in the following table:

Event		$\mathcal{O}$	$\mathcal{P}$
Coordinates in Σ:	ct	0	cT
	z	0	uT
Coordinates in Σ':	ct'	0	cT'
	z'	0	0

Lorentz tranform coordinates of $\mathcal{P}$ in Σ to the time coordinate in Σ' using equation (7.5)

4 The transformation from Σ to Σ' gives

$$cT' = \gamma_u(cT - \beta_u uT) = \gamma_u cT(1 - \beta_u^2) = cT/\gamma_u.$$

5 Thus $T = \gamma_u T'$ and since γ is always greater than 1, $T > T'$, the clock C_0' ticks away time more slowly than do the stationary synchronised clocks that it passes as it moves. This is time dilation, the slow running of a clock moving past stationary synchronised clocks.

z'-axes so we consider this, as we have previously, as a problem in time and one space dimension alone. Study Figure 7.9 which demonstrates the property of time dilation. A time interval T' registered on a moving clock is less than that, T, registered on the synchronised, stationary clocks that it passes at speed u. Formally

$$T' = T/\gamma_u = T\sqrt{1 - u^2/c^2}. \tag{7.9}$$

There are some comments that we have to make about this result.

1   There is a thought that this is not a relativistically correct result because it distinguishes an inertial frame Σ that is absolutely stationary. That is incorrect. Note that in our derivation (Figure 7.9) there are three clocks involved: C_0' in Σ' and C_0 plus C_1 in Σ. So the situation is not symmetric between these two frames. If we take the situation that is really symmetric to the one we have just described we have to have a clock C_0 stationary in Σ that is moving with speed u along the negative z'-axis in Σ'. It passes from clock C_0' to a clock C_1' and again the 'moving' clock will appear to be ticking more slowly than the clocks it passes that are stationary in Σ'. The two situations are symmetric, the time dilation is symmetric and neither frame is privileged as being absolutely stationary. The principles of special relativity are not violated.

2   There is a very straightforward and simple way of obtaining this time dilation result. It is given as a problem for the reader to complete (Problem 7.3).

3   The partial derivative

$$\left.\frac{\partial t}{\partial t'}\right|_{z'}$$

is the ratio of infinitesimal time interval between two events as measured in Σ to the corresponding interval in Σ', obtained under conditions in which the latter interval is measured at constant z'. That is just the conditions (clock C_0' stationary in Σ') that revealed the time dilation formula for finite time intervals. Now from equation (7.5) $ct = \gamma_u(ct' + \beta_u z')$, therefore

$$\left.\frac{\partial t}{\partial t'}\right|_{z'} = \gamma_u. \tag{7.10}$$

This is the same result for infinitesimal time intervals as was found for finite time intervals (equation 7.9) and derived in Figure 7.9.

4   The matter of the experimental confirmation of time dilation is discussed in the next section.

**7.11 The experimental tests of time dilation

In the early days of special relativity there were no clocks that could be moved at speeds approaching the speed of light. Direct, precise tests of time dilation only became possible with the development of experimental elementary particle physics. This provided two things:

1   The discovery of many kinds of unstable particles, each kind having a charac-
    teristic mean life before decay. Such a mean life provides a natural clock,
    associated with what is essentially a point mass.
2   The ability to produce and detect such particles moving at speeds, relative to
    a laboratory, that are close to the speed of light.

The Radioactive Decay Law states that if there are $N(0)$ unstable particles of a
specific kind at time $t = 0$, then at time t the number surviving undecayed, $N(t)$,
is given by

$$N(t) = N(0) \exp(-t/t_{\mathrm{D}}), \tag{7.11}$$

where t_{D} is the mean life of the particular kind of particle being considered. This
is, of course, a statistical statement. It is impossible to predict the lifetime of an
individual particle. And N is an integer so cannot follow the smooth, time-wise
falling exponential. Thus the law holds only in the limit of an infinite number of
particles. None the less many mean lives are known with considerable precision
and frequently large enough samples of particles could be observed for the law to
be an excellent description of the time decay of these samples.

The law applies in the rest frame of the particle. What happens if the particle
is moving? We consider a large group of a specific kind of particle moving with
speed u in the same direction in a laboratory. Non-relativistically, if $N(0)$ is the
number in the group at one point on its path, then the number surviving after a
distance l, $N(l)$, is given by

$$N(l) = N(0) \exp(-l/u t_{\mathrm{D}}).$$

The mean distance travelled before decay is $u t_{\mathrm{D}}$. This applies only if $u \ll c$. For
relativistic speeds the moving clock runs slow compared to the clocks stationary
in the laboratory. In particular, the natural tick time t_{D} of the particles is dilated
as observed by the laboratory clocks. The dilating factor is $\gamma_u (= 1/\sqrt{1 - \beta_u^2}$,
$\beta_u = u/c)$. In the laboratory the mean life t_{lab} is given by

$$t_{\mathrm{lab}} = \gamma_u t_{\mathrm{D}}. \tag{7.12}$$

The distance survival equation becomes

$$N(l) = N(0) \exp(-l/\gamma_u u t_{\mathrm{D}}).$$

The mean distance λ moved before decay is given by

$$\lambda = \gamma_u u t_{\mathrm{D}} = \gamma_u \beta_u c t_{\mathrm{D}}. \tag{7.13}$$

For the particles $\gamma_u \beta_u = P/Mc$ (Table 2.2) so that

$$\lambda = (Pc/Mc^2) c t_{\mathrm{D}}. \tag{7.14}$$

Thus the mean distance travelled by the members of a sample of these unstable
particles before decay is proportional to their common momentum. Alternatively,

we can say that the probability a single particle will move a distance λ without decay is e^{-1}.

Here is an example of the numbers involved. Consider the muon (see Figure 3.1):

$$M_\mu c^2 = 105.66 \, \text{MeV}, \quad t_D \equiv t_\mu = 2.1970 \times 10^{-6} \, \text{s}, \quad ct_\mu = 658.65 \, \text{m}.$$

For a muon of momentum $Pc = 1000 \, \text{MeV}$ we have

$$\gamma_u \beta_u = 1000/105.66 = 9.4643, \quad \gamma_u = \sqrt{1 + (\gamma_u \beta_u)^2} = 9.5170,$$

$$\beta_u = 0.99446.$$

Without time dilation we would expect such muons to survive a mean distance of $\beta_u ct_\mu$, about 655 metres. Einstein tells us to expect the distance to be $\gamma_u \beta_u ct_\mu$ which is 6.234 km.

Muons were the subject of one of the first tests of time dilation. This particle was discovered in 1936 in the cosmic radiation at ground level. They are generated by the in-flight decay of energetic, charged pions ($\pi \rightarrow \mu + \nu$, Section 3.3) produced by primary cosmic radiation in the upper layers of the atmosphere. Without time dilation the chance of these muons reaching ground level from production at a height of more than 10 km is negligible. The existence of time dilation changes the situation and a substantial number of energetic muons reach ground level. The early test[3] measured the number of muons at a height of 3240 and 1616 m. The apparatus was designed to detect muons at the higher position in a certain energy range and at the lower position the same energy range of muons by allowing for the energy loss experienced by these muons piercing the layer of air between the two heights. The numbers found at the lower level were consistent with losses due to a time dilated decay, but not with the absence of dilation.

A weakness of the cosmic ray muon test was that the muons descending through the atmosphere were undergoing collisions with electrons bound in air molecules. Each collision causes a loss of energy from the muon. Each loss is small, but there are many collisions. Thus the muons are gradually losing speed and they do not remain at rest in one inertial frame. This is not quite the situation that applied in our derivation of the time dilation. However, integration of the effect over many changes of inertial frame can be justified theoretically.

There is an example of a similar situation in which the moving, decaying particles remain in a vacuum and are not subject to the energy loss deceleration. In 1978, CERN commissioned a 280 GeV muon beam as a part of the research facilities associated with the 400 GeV Super Proton Synchrotron (SPS).

The muons were produced by the in-flight decay of charged pions in a beam having a momentum close to 300 GeV and travelling in a vacuum pipe. The distance in which decay occurs usefully is 600 m. The majority of muons are produced moving very close to the direction of their parent pion and can be considered as remaining a component of the beam. Without time dilation all the pions would

---

3 Rossi, B. and Hall, D. B., *Physical review* **59**, 1941, pp. 223–8.

decay in this distance. With time dilation about 3.5% of the pions are expected to decay (see Problem 7.4). After the 600 m, the now mixed beam (surviving pions plus muons) passes through a filter (10 m of beryllium) which absorbs the remaining pions and transmits almost unaffected the muons. The final muon beam intensity had a value expected from the design that allowed for time dilation throughout. Without time dilation the final muon beam would have had an intensity nearly 30 times greater than that observed. (Of course, this simplified description misses many physics and engineering issues. For example, what happens to the neutrinos also produced in the decay of the pions? They have no significant effect on anything, but why? We leave it to the reader to discover the answer.)

However, even with the vacuum, the decaying pions do not remain securely in one inertial frame. Throughout the 600 m the beam is kept from diverging by magnetic lenses so that a pion straying from beam centre line, as most do, experiences a restoring $\mathbf{v} \times \mathbf{B}$ force and smoothly changes inertial frames! This is a very different circumstance from the deceleration of cosmic ray muons and yet the calculation using the simple formula for the time dilation is still correct. This suggests that acceleration does not affect the clocks provided by particle decay.

That example is not as quantitative a test as is desirable. In the next section we shall encounter a verification of the time dilation formula with a precision of 0.2% but with the complication that the moving clock is confined to an accelerating frame. The best measurements of time dilation of an unaccelerated clock have been performed using the particles having the label K^0. (These electrically neutral particles do not suffer either gradual kinetic energy loss when moving through materials as do muons descending though the atmosphere, or the $\mathbf{v} \times \mathbf{B}$ force experienced by pions in magnetic lenses.) In Figure 3.4 the particle that invisibly moved through liquid hydrogen from vertex 0 to vertex 1 was one of these. Taking $\gamma = E_K / M_K c^2$ the mean lifetime in the laboratory before decay has been measured, in different experiments, at $\gamma = 1.63$, 15.2, and 20.7. The agreement with the formula (equation (7.12)) was within the error of 0.7%.

**7.12   The travelling twin

The result of this story is sometimes called *The Twin Paradox* or *The Clock Paradox*. This is misleading: there is no paradox. The result is understood and, as we shall see, has been verified.

Twins Castor and Pollux agree that Castor will stay at home during the time Pollux makes a there-and-back journey to the star alpha Centauri. This star is known to be at a distance of four light years and Pollux makes the journey at a steady speed of $u = 0.8\,c$ relative to the inertial frame of stay-at-home Castor. To the latter, Pollux is away for a time

$$2 \times \text{distance to alpha Centauri} \div \text{Pollux's speed} = 2 \times 4/0.8 = 10 \text{ years.}$$

The time dilation factor for Pollux is $\gamma^{-1} = \sqrt{1 - u^2/c^2} = 0.6$ and his clocks, real and biological, tick time away more slowly by this factor relative to clocks stationary in Castor's inertial frame. Thus while Castor has aged 10 years during the period of Pollux's absence, Pollux has only aged six years during his journey. Reunited the twins discover Castor is now four years older than his brother Pollux (see Problem 7.6).

Now we can make some comments on this result, particularly in the light of some of the criticisms made by those who found it difficult to accept this result of time dilation.

- How do we know that our biological clocks are subject to the same time dilation as our physical clocks? (1) Biological effects are a manifestation of physical and chemical laws which are subject to Einstein's principles of relativity. (2) If biological ageing was the same as clock time ageing in one inertial frame but different in others it would be possible to identify a privileged inertial frame, something those principles forbid. We have to accept that biological ageing is dilated.
- Castor sees Pollux depart and return. Equally Pollux sees Castor depart and return. Is it not so that the situation is symmetric and the twins should age equally? No! The situation is not symmetric. Pollux does not remain in a unique inertial frame. He experiences a program of accelerations and decelerations not experienced by Castor. Near alpha Centauri these must lead to a change of inertial frame for Pollux from one moving away from Castor to one moving towards him. Otherwise, how will Pollux return to Castor? The latter remains comfortably in one and the same inertial frame throughout Pollux's voyage.
- There is a question as to whether the difference in ageing might be wholly or partially caused by the periods of acceleration and deceleration experienced by Pollux at the beginning and end of each of his outgoing and incoming journeys. However, suppose Castor also makes a trip lasting one year to Pollux's ten but with the same program of accelerations and decelerations at the begining and end of each leg of their respective trips. If accelerations specifically contributed to the rate of ageing then that particular effect would cancel when considering the difference in ageing. The time dilation effect would still apply to the difference in the length of time spent on unaccelerated movement. The twins would still find an age difference when they were reunited.
- Of course all but the gentlest accelerations and decelerations can have a disastrous effect on biological systems. The twins must organise their journeys with this in mind if they are to avoid injuries.

However, the existence of the accelerations has to be addressed. Special relativity allows a treatment of acceleration, assuming that an accelerated clock is an ideal clock (Section 5.2). Then a finite time period on the accelerated clock can be related to that of synchronised clocks stationary in the inertial frame in which the velocity of the ideal clock is defined. This is done by appropriate integration of equation (7.10). Thus Pollux's (ideal) clock change may be properly related to that of stay-at-home Castor. Note that the calculation depends on Castor remaining stationary in the same inertial frame throughout Pollux's travels.

Suppose the problem is approached from the point of view of traveller Pollux. If we define a frame of reference in which he remains stationary, then we cannot use special relativity to describe the movements of Castor within this frame. The reason is that this frame of reference is not inertial, and the proper treatment requires the techniques of general relativity.

The calculations made from these two points of view agree and, if the periods of acceleration or deceleration are small compared to the total time spent on the journey, the result of a simple calculation of the kind that was made at the beginning of this section will be nearly correct.

The journal *Nature* was the forum for a controversy about time dilation and the clock paradox in 1956. The matter reappeared, again in *Nature*, in 1974–5, sustained by one fierce critic of special relativity. It is interesting, therefore, to note that the same journal published, in 1977, a paper describing what was then the most precise confirmation of time dilation and of the twin paradox.[4] CERN had built a Muon Storage Ring with the primary objective of measuring the anomalous part of the muon magnetic moment to a precision of 1 part in 10^6. The precision confirmation of time dilation and of the clock paradox was a bonus. The ring allowed a measurement of the mean life of samples of moving muons, momentum $3.094\,\text{GeV}/c$ ($\gamma = E/Mc^2 = 29.30$), maintained in a closed, circular path (in a vacuum) by a laboratory magnet. The positive muon's known 'at rest' mean life of $2.19703\,\mu\text{s}$ was observed to be dilated to $64.42 \pm 0.06\,\mu\text{s}$ which agrees with special relativity within the experimental accuracy of 0.2%.

There are some points that this result raises.

1   The moving muons are subjected to an acceleration of $10^{19}\,\text{m s}^{-2}$ ($10^{18} \times$ the acceleration due to gravity at the surface of the Earth). From the view point of the experimenter stationary in the laboratory the dilation is that predicted by special relativity and the clock defined by the mean lifetime of the muon is ideal.

2   The experiment simulates the outward and return journey of the twin paradox. Thus the experiment has shown that a sample of muons stationary at a point on the circular path age $1.468 \times 10^{-7}\,\text{s}$ during the time a second sample moving with speed $0.99942c$ has been once around the circular path and aged $5.01 \times 10^{-9}\,\text{s}$. It is the fraction of muons decaying in each sample that measures the ageing.

3   Accelerations are involved so what does general relativity say about this situation? In the non-inertial rotating reference frame in which the circulating muons are at rest, an observer, also at rest, would discover that these muons are decaying more slowly than a sample of muons stationed on the axis of rotation. This observer ascribes this effect to the potential of a gravitational field that is equivalent to the centrifugal force that is experienced away from

4 Bailey, J., *et al.* Measurements of relativistic time dilation for positive and negative muons in a circular orbit. *Nature* **268**, 1977, pp. 301–5.

the axis. This potential is zero for muons stationed on the axis but negative for the muons stationed a distance from the axis of rotation. General relativity predicts that the more negative this potential is at the position of a clock, the slower it ticks. Thus the muons not on the axis decay more slowly than do those on the axis. The result of this analysis is the same as that provided by special relativity.[5]

*7.13   Proper time

We have discussed time dilation. This has involved a moving clock passing stationary clocks, and appearing to run slow as it did so. Let us now consider a particle moving in an inertial frame Σ and suppose this particle has an ideal clock. If the particle is moving uniformly with speed v and it passes from one event to a second event in Σ which is separated from the first by a time interval Δt measured on synchronised clocks in Σ, then the time interval shown on the particle's clock, $\Delta\tau$, is given by

$$\Delta\tau = \Delta t \sqrt{1 - v^2/c^2}.$$

We can reduce the finite intervals to infinitesimal intervals and put

$$d\tau = dt \sqrt{1 - v^2/c^2}, \qquad (7.15)$$

where v is now the instantaneous speed of the particle in Σ. Since we accept that acceleration has no effect on an ideal clock, we are able to integrate over a part of the particle's world line, even if v is varying, to obtain the corresponding finite interval of τ:

$$\tau = \int dt \sqrt{1 - v^2/c^2}. \qquad (7.16)$$

The quantity τ is called the **proper time**. We are free to set a zero of τ at any convenient event on the world line of the particle. In particular, we are now free of the need to imagine a clock 'attached' to the particle and can work in terms of the particle's proper time.

The name *proper time* for the time in the rest frame of a particle reminds us of the *proper length* as the length of a rod in the inertial frame in which it is at rest. It also suggests that the *rest mass* of a particle might more correctly be called the *proper mass*, although that is not usual.

The most important property of the proper time is that it is a relativistic invariant in the following sense. In Section 7.6 we introduced the invariant infinitesimal interval between two events:

$$ds^2 = dct^2 - dx^2 - dy^2 - dz^2.$$

---

5 Møller, C., *The theory of relativity*, 2nd edn, Oxford: Clarendon Press, 1972, pp. 280–1, 292–6.

In the case that these two events are on the world line (Section 7.5) of our particle we have in its rest frame

$$dx' = dy' = dz' = 0 \quad \text{and} \quad dct = dc\tau.$$

Therefore $|dc\tau| = |ds|$ and $d\tau$ is an invariant. From whatever inertial frame we choose to observe the particle, the change in τ between two events on its world line is always the same.

*7.14   Time dilation and the Lorentz–FitzGerald contraction

Consider a clock C′ stationary in the inertial frame Σ' that is moving so that the C′ moves with speed u along the z-axis of inertial frame Σ (the standard arrangement). In a time interval t recorded on synchronised clocks in Σ, C′ will move a distance ut along z in Σ and record a time interval $t' = t\sqrt{1 - u^2/c^2}$. An observer in Σ' might ask 'what is the distance the origin of Σ has moved in Σ' in the period t'?' The answer cannot be ut because if u was close to c that observer might deduce that the speed of Σ in Σ' was ut/t' which can be greater than c and is therefore impossible. Of course, the answer is that to the observer in Σ' the distance Σ has moved in Σ' in time t' is ut contracted by a factor $\sqrt{1 - u^2/c^2}$. Then the speed the Σ' observer deduces is

$$\left(ut\sqrt{1 - u^2/c^2}\right) \Big/ \left(t\sqrt{1 - u^2/c^2}\right) = u.$$

Thus

1   the observer in Σ deduces that Σ' (and C′) are moving with speed u in the direction of the z-axis, and
2   the observer in Σ' deduces that Σ is moving with speed u but in the direction of the negative z'-axis.

Revisit the description of the CERN muon beam (Section 7.11). In the laboratory the pions have a free path of 600 m. To an observer in an inertial frame in which the pions are at rest this distance is 27.9 cm and features of the laboratory are passing at a speed that is $0.999999892c$. Problem 7.4 asks you to confirm these numbers.

7.15   The transformation of velocities

Consider a particle moving in the Σ' inertial frame so that at time t' its space coordinates are x', y' and z'. Then the components of its velocity $\mathbf{v}'$ are

$$v'_{x'} = \frac{dx'}{dt'}, \quad v'_{y'} = \frac{dy'}{dt'}, \quad \text{and} \quad v'_{z'} = \frac{dz'}{dt'}.$$

What are the components of this velocity $\mathbf{v}$ observed in Σ? Again we define

$$v_x = \frac{dx}{dt}, \quad v_y = \frac{dy}{dt}, \quad \text{and} \quad v_z = \frac{dz}{dt}.$$

We can differentiate the equations (7.2) of the Lorentz transformations of the space–time coordinates to give

$$dt = \gamma_u (dt' + [u/c^2] dz'),$$

$$dx = dx',$$

$$dy = dy',$$

$$dz = \gamma_u (dz' + u dt').$$

Then

$$v_x = \frac{dx}{dt} = \frac{dx'}{\gamma_u (dt' + [u/c^2] dz')},$$

therefore

$$v_x = \frac{v_{x'}'}{\gamma_u (1 + u v_{z'}'/c^2)}. \tag{7.17}$$

Similarly

$$v_y = \frac{v_{y'}'}{\gamma_u (1 + u v_{z'}'/c^2)} \tag{7.18}$$

and

$$v_z = \frac{v_{z'}' + u}{(1 + u v_{z'}'/c^2)}. \tag{7.19}$$

These equations have some properties that should be noted:

1   They give the components (**v**) of a velocity as observed in a frame Σ in terms of the components (**v'**) observed in a frame Σ'. To obtain the reverse transformation make the following changes in equations (7.17)–(7.19):

$$\mathbf{v} \to \mathbf{v'}, \quad \mathbf{v'} \to \mathbf{v}, \quad \text{and} \quad u \to -u.$$

2   In the limit where all velocities are small compared to c, or when $c \to \infty$, these equations reduce to those applying to a Galilean transformation of velocity components, equations (5.6)–(5.8).

3   The two inertial frames must have a relative velocity $u < c$. Then consider the case where $v_x' = 0$, $v_y' = 0$ but $v_z' \to c$ then v_z remains less than c. So whatever increment of speed is added to v_z' by the boost from Σ' to Σ, the result never reaches c. If $v_z' = c$ then $v_z = c$ which is the expected result given the principle that the speed of light is the same in all inertial frames. Or, to put it another way, given that neither v_z' nor u is greater than c and if one or both are less than c, then so is v_z. Thus, as we expect, c remains a limiting speed.

We stated in Section 4.3 that we wished to give no particular importance to velocity as a variable in special relativity. One of the reasons is that the equations (7.17)–(7.19) do not exhibit the simple structure of the transforming equations for a Lorentz boost, as in equations (6.5) and (7.5). It is not easy, nor is it necessary, to memorise these equations. There is another way to transform velocities once the Lorentz transformation is properly understood. Consider the four quantities

$$\gamma_v c, \quad \gamma_v v_x, \quad \gamma_v v_y, \quad \gamma_v v_z,$$

where

$$\gamma_v = 1/\sqrt{1 - v^2/c^2}$$

and

$$v^2 = v_x^2 + v_y^2 + v_z^2.$$

We shall begin to view such four quantities as constituting the components of a *four-velocity*. They transform under Lorentz transformations as do the four quantities

$$ct, \quad x, \quad y, \quad z.$$

(We prove this in Section 8.10.) Problem 7.5 asks the reader to confirm that using this property gives identical results to equations (7.17)–(7.19).

We shall discuss the appearance of four-component quantities all transforming identically in Chapter 8.

7.16   The composition of velocities

This title is conventionally applied to what is essentially a transformation of velocity from one inertial frame to another. However, we are about to look at an example of problems that require a careful choice of transformation.

Consider a particle (1) moving with velocity **u**. A second particle (2) is moving with velocity **v**. The problem is: What is the velocity of the second particle as observed from a frame in which the first is at rest? We can call this the **relative velocity**. It is not $\mathbf{v} - \mathbf{u}$! That is the Newtonian result, and called the *mutual velocity*. To find the relative velocity, look at Figure 7.10. We organise an inertial frame Σ in which **u** is parallel to the z-axis. Then, using the standard arrangement we organise a frame Σ' that has a speed u along the z-axis in Σ (Figure 7.10(a)). In Σ' particle 1 is at rest. The transformation of the components of the velocity (**v**) in Σ to their values (**v**') in Σ' (Figure 7.10(b)) will give the required result. Before we do that we note that, as formulated, there is a freedom to orient the x- and y-axes of Σ any way we wish with respect to rotations around the z-axis. This is not important in the physics of this problem. Follow a procedure as in the previous section; it is the same except the velocity transformation is from Σ to Σ'

(a)

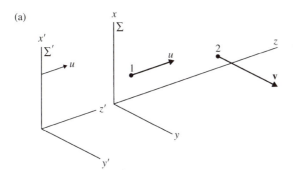

(b)

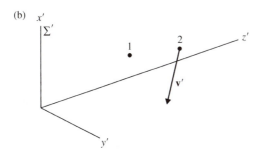

Figure 7.10 The arrangement of axes for two inertial frames used to find the velocity
of particle 2 with respect to particle 1. In (a) the axes of Σ are oriented so
that **u** (velocity of 1) is parallel to the z-axis. Also in (a) the orientation
of Σ' is as usual. In (b) we see that particle 1 is stationary in Σ' and the
velocity of 2 (**v**') is now different from its value in Σ. Therefore **v**' is the
velocity of 2 as observed from the rest frame of 1.

instead of the other way! So if the components of **v** are v_x, v_y, and v_z, then the
components of **v**' in Σ' are

$$v'_{x'} = \frac{v_x}{\gamma_u(1 - uv_z/c^2)},$$ (7.20)

$$v'_{y'} = \frac{v_y}{\gamma_u(1 - uv_z/c^2)},$$ (7.21)

and

$$v'_{z'} = \frac{v_z - u}{(1 - uv_z/c^2)}.$$ (7.22)

This result is the relative velocity of particle 2 with respect to particle 1 given
the choice of orientation of the axes involved. The reverse relative velocity is the
negative of this.

Therefore, finding the relative velocity between two particles is a matter of
defining two inertial frames and making a velocity transformation in the correct

direction. If there is any doubt about the sign that goes with u take the limit $c \to \infty$ and check that the expected Newtonian result is found.

A few moments' thought will show that this method will in general require a rather complicated rotation of axes in ordinary space to bring the velocity $\mathbf{u}$ parallel to the z-axis of Σ. The same rotation must also be applied to the velocity $\mathbf{v}$. If the components of $\mathbf{u}$ are known in some reference frame then it is possible to make a Lorentz transformation along the direction of $\mathbf{u}$ without any rotations. The matrix that does this is given in equation (8.23), Section 8.7. The speed associated with the required transformation is u and correctly applied it takes the system into an inertial frame in which particle 1 is at rest. The same matrix applied to the four-velocity of particle 2, namely $(\gamma_v c, \gamma_v \mathbf{v})$, will give its velocity in this new frame. And that is the four-velocity $(\gamma_w c, \gamma_w \mathbf{w})$, where $\mathbf{w}$ is the three-velocity of 2 relative to 1 at rest (see Problem 8.9). However, note that the components of $\mathbf{w}$ will now be given with respect to coordinate axes that are not the same as those (Σ') used to derive equations (7.20)–(7.22).

+7.17   Minkowski map

Three years after the publication of Einstein's first paper on special relativity, the mathematician Hermann Minkowski (1864–1909) proposed a geometrical approach to special relativity. He emphasised the space–time of a four-dimensional world in place of a three-dimensional space plus time. Some quantities that are represented by vectors with three components in the Newtonian world have to be found a fourth component which produces a quantity, a *four-vector*, in the Minkowski world (Section 8.10). To us there is nothing astonishing in this since that is just how we have been treating the time–space coordinates of events and the energy–momentum of a particle. However, Minkowski's early emphasis on this aspect led the way to the use of tensors not only in special relativity but also, in particular, in general relativity. We shall look at some of the uses of four-vectors (rank-1 tensors!) in the next chapter (Sections 8.9 and 8.10).

In Section 7.5 we described what was Minkowski's idea, that of a *world line*. Figure 7.3 shows two examples of world lines mapped onto the ct–z plane. In Figures 7.4 and 7.5 we indicated how the world lines of photons converging to an event or diverging from the same event defined regions of space–time such as the *past* and the *future* to such an event (Section 7.5). Minkowski developed the use of such maps so that it was possible to represent a world line or an event on one map and read off coordinates in two different inertial frames simultaneously. Figure 7.11 outlines the steps in the construction of such a map for the two inertial frames Σ and Σ' of the standard arrangement. As in Figure 7.3, our map will have two axes for each inertial frame, ct, z for Σ and ct', z' for Σ'. Thus it can only map the events that occur on the $z(z')$-axis or the world lines of particles that move along the $z(z')$-axis. The completed map is shown in Figure 7.11(d). It satisfies the requirements of giving the coordinates of any event both in Σ and in Σ'. Note that the coordinate axes ct' and z' are not orthogonal. Thus the event $\mathcal{M}$ has coordinates $ct = 0.68$, $z = 0.68$, $ct' = 0.5$ and $z' = 0.5$, all in metres.

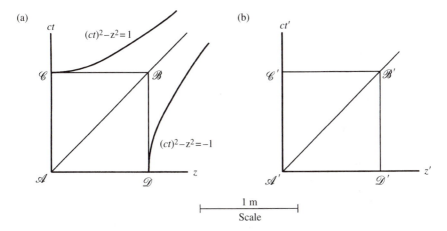

Scale

Figure 7.11 The construction of a Minkowski map for our usual arrangement of two inertial frames, Σ and Σ', for events that occur on the $z(z')$-axis. The scale is that shown. The unit of all coordinates is a metre.

i Diagram (a) is a map for the Σ frame. The world line of a particle stationed at the origin of Σ is along the ct-axis. At event $\mathcal{A}$ the particle's coordinates are $(ct, z)_{\mathcal{A}} = (0, 0)$ and after $1/c$ seconds the particle is at the event $\mathcal{C}$ on the map where $(ct, z)_{\mathcal{C}} = (1, 0)$. The world line for a photon emitted along the z-axis at event $\mathcal{A}$ is the line $\mathcal{AB}$ that lies at equal angles to the ct- and z-axes. The event $\mathcal{B}$ on the photon world line has coordinates $(ct, z)_{\mathcal{B}} = (1, 1)$. A particle at rest on the z-axis would have a world line that is parallel to the ct-axis. If it is at rest at $z = 1$, then event $\mathcal{D}$ with coordinates $(ct, z)_{\mathcal{D}} = (0, 1)$ and event $\mathcal{B}$ are on this particle's world line. Diagram (b) is a map for the Σ' frame. The origins of Σ and Σ' coincide at event $\mathcal{A}'$ ($\mathcal{A}$ on diagram (a)). Now the events $\mathcal{A}'\mathcal{C}'\mathcal{B}'\mathcal{D}'$ play the same role in (b) as the events $\mathcal{ABCD}$ played in (a). However, of the two sets of events, only $\mathcal{A}$ and $\mathcal{A}'$ are the same event.

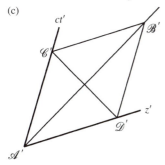

ii Distort (b) linearly, stretching it along $\mathcal{A}'\mathcal{B}'$ and squeezing it along $\mathcal{C}'\mathcal{D}'$. The result is that the square $\mathcal{A}'\mathcal{C}'\mathcal{B}'\mathcal{D}'$ becomes a rhombus shown in (c). The whole is, in its information content, identical to (b). The axes are now skew and for the moment the scale along each axis remains the same as in (a). The coordinates of event $\mathcal{B}'$ are still $ct' = 1$ and $z' = 1$.

(Continued)

Figure 7.11 (Continued)

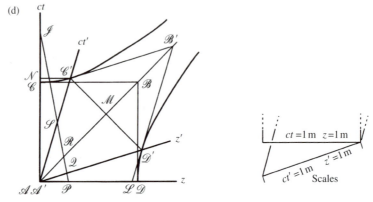

iii Now superimpose the map (c) on map (a) making $\mathcal{A}'$ coincide with $\mathcal{A}$ and the direction $\mathcal{A}'\mathcal{B}'$ with the direction $\mathcal{A}\mathcal{B}$. Now we have to make some adjustments to angles and scales of (c) with the objective that a world line drawn on (a) is also the same world line in (c). The result is diagram (d). The adjustments that need to be made are now described.

iv The frame Σ' is moving in the usual way in Σ and the world line of the origin of Σ' is along $\mathcal{A}'\mathcal{C}'$. It moves with speed $u = \beta_u c$ and we can use the Lorentz transformation to fix angles. Using the reciprocal transformation of equation (7.5) we see that the line $ct' = 0$ ($\mathcal{A}'\mathcal{D}'$) is $ct - \beta_u z = 0$. It follows that the angles $\angle\mathcal{C}\mathcal{A}\mathcal{C}'$ and $\angle\mathcal{D}\mathcal{A}\mathcal{D}'$ are both $\tan^{-1}\beta_u$. As drawn, this angle is 16.7° and $\beta_u = 0.3$. Any other angle on the map which is not a right angle follows from that value.

v We can now adjust the scale of the rhombic part of the map using the invariance of $(ct)^2 - z^2 = (ct')^2 - z'^2$. On (a) and (d) branches of the two hyperbolae $(ct)^2 - z^2 = \pm 1$ are drawn. The ct' axis, that is the line $z' = 0$, must intersect the $+1$ hyperbola at $ct' = 1$. The labelling on (d) has made $\mathcal{C}'$ that event. So the length $\mathcal{A}'\mathcal{C}'$ on (d) represents 1 m of ct' in Σ'. Similarly the line $ct' = 0$ intersects the -1 hyperbola at $z' = 1$. The labelling makes $\mathcal{D}'$ that event. And the length $\mathcal{A}'\mathcal{D}'(=\mathcal{A}'\mathcal{C}')$ on the map (d) represents one metre of z' in Σ'. Since $\mathcal{A}'\mathcal{D}' > \mathcal{A}\mathcal{D}$ and $\mathcal{A}'\mathcal{C}' > \mathcal{A}\mathcal{C}$ the rhombus is expanded but not distorted. Event $\mathcal{B}'$ still has $(ct', z')_{\mathcal{B}'} = (1, 1)$. During this scaling the distances on the (a) part of the map remain the same: $\mathcal{A}\mathcal{C} = \mathcal{A}\mathcal{D} = 1$.

We can now state these facts and others about the map (Figure 7.11(d)):

a The *scale* of the map relates a length on the ct- or z-axis or on the ct'- or z'-axis to a distance in time or space.

b The scale of the ct- and z-axes is different from the scale of the ct' and z'-axes.

c On the map:

$$\mathcal{A}'\mathcal{C}' = \mathcal{A}'\mathcal{D}' = \sqrt{\frac{1 + \beta_u^2}{1 - \beta_u^2}}\,\mathcal{A}\mathcal{C} \quad (=1.094\ \mathcal{A}\mathcal{C}\ \text{for this map}),$$

$$\mathcal{A}\mathcal{C} = \mathcal{A}\mathcal{D}.$$

d As drawn, a map length on the ct- or z-axis equal to $\mathcal{A}\mathcal{C}$ is equivalent to 1 m in Σ.

e As drawn, a map length on the ct'- or z'-axis equal to $\mathscr{A}'\mathscr{C}'$ is equivalent to 1 m in Σ'.

f $\angle\,\mathscr{C}\mathscr{A}'\mathscr{C}'= \angle\,\mathscr{D}\mathscr{A}'\mathscr{D}'= \tan^{-1}\beta_u (= 16.7° = \tan^{-1}0.3$ on this map).

g The map extends into all four quadrants. All the axes ct, z, ct', z' extend to negative values if $\mathscr{A}$ ($\mathscr{A}'$) is the origin for all. The photon world line may be extrapolated back into the third quadrant if we need to think about the photon approaching the event $\mathscr{A}$ from negative z. Or constructed in the second and fourth quadrants if we wish to represent a photon moving in the opposite direction.

h Remember a world line on this map represents a particle or photon moving along the z-axis in Σ and the z'-axis in Σ'. The particle cannot have a speed greater than c. This means that with respect to the ct- and z-axes the tangent to the world line must be at an angle less than $\pm\pi/4$ from the ct-axis. If this is the case then the particle's speed along the z'-axis will also be less than c.

i We have drawn a simple world line. It represents a particle that starts, at $ct = 0$ (event $\mathscr{P}$), to move along the z-axis from $z = 0.3$ m with speed $dz/dt = -0.2c$. Thus it is heading towards the origin of Σ. At the event $\mathscr{Q}$ it meets a point on the z'-axis of Σ' at which the local clock reads $ct' = 0$. At the event $\mathscr{R}$ it meets the photon that left the event $\mathscr{A}$ ($\mathscr{A}'$) when the origins of Σ and Σ' coincided, and travelled along the z-axis. At $\mathscr{S}$ it meets the origin of Σ' as that point advances along the z-axis. At $\mathscr{T}$ it reaches the origin of Σ.

+7.18 The Lorentz–FitzGerald contraction and time dilation on the Minkowski map

Consider a rod of length 1 m at rest in Σ' with one end at the origin of Σ'. There is a world line associated with each end of the rod. One world line (Figure 7.11(d)) coincides with the ct'-axis and at $ct' = 0$ that end is, by definition, at $\mathscr{A}'$. The world line of the other end is a straight line parallel to the ct'-axis. Since $\mathscr{A}'\mathscr{D}'$ corresponds to one metre in Σ' that world line passes through event $\mathscr{D}'$ when $ct' = 0$ and event $\mathscr{B}'$ at a later time. We extend $\mathscr{B}'\mathscr{D}'$ to intersect the z-axis at event $\mathscr{L}$. The world line of this end of the rod is $\mathscr{L}\mathscr{B}'$. Thus the map length $\mathscr{A}'\mathscr{D}'$ represents the rod length of 1 m when it is at rest in Σ' at $ct' = 0$. Consider now the length of the rod as observed in Σ. We have to find the position of the ends of the rod when the clocks on the z-axis in Σ read the same time. A convenient time is $ct = 0$, and the corresponding events are $\mathscr{A}$ and $\mathscr{L}$. Clearly the length $\mathscr{A}\mathscr{L}$ represents the measured length of the rod and that it is less than one metre. And there we have the Lorentz–FitzGerald contraction. A careful application of geometry, including to the distance scales, will show that $\mathscr{A}\mathscr{L}$ correspond to $\sqrt{1-\beta_u^2}$ metres in Σ, as expected.

Consider a particle at rest at the origin of Σ'. At event $\mathscr{A}'$ we set its proper time τ to 0. Its world line coincides with the ct'-axis. Therefore when $c\tau = 1$ m the world line reaches $\mathscr{C}'$. At this event the ct coordinate corresponds to the length $\mathscr{A}\mathscr{N}$. This is clearly greater than 1 m, hence time dilation. Another careful piece of geometry shows $c\tau = ct\sqrt{1-\beta_u^2}$ at event $\mathscr{C}'$, as expected.

That was a long and arduous route to find time dilation and the Lorentz–FitzGerald contraction. However, understanding the Minkowski map adds to a general understanding of the Lorentz transformations.

*7.19   Summary

This ends the chapter of the two challenges to intuition that are posed by special relativity, namely time dilation and the Lorentz–FitzGerald contraction. In the problems at the end of the chapter we have covered both these subjects. In fact, many associated paradoxes have been posed in connection with special relativity. For example, what is happening to speeding poles 5 m long trapped inside barns 4 m long? All these are challenges to which an appreciation of simultaneity is the key. Problem 7.15 is set in a manner that gives some clues how to think about these paradoxes. Problem 9.10 asks for a solution to a second paradox: the clue is in the host chapter.

Problems

Wherever the inertial frames Σ and Σ' are mentioned, the standard arrangement of Figure 7.1 is implied.

7.1   a   The Lorentz transformation of equation (7.2) changes the time–space coordinates $(t', \mathbf{r}')$ in Σ' of an event to the time–space coordinates $(t, \mathbf{r})$ in Σ of the same event. Show that

$$(ct')^2 - \mathbf{r}' \cdot \mathbf{r}' = (ct)^2 - \mathbf{r} \cdot \mathbf{r}.$$

b   Confirm that two successive Lorentz transformations that go from Σ' to Σ and then back to Σ' leave the time–space coordinates of an event unchanged.

7.2   Consider two events for which the interval between them is time-like (Section 7.6). Show that the time-ordering of these two events is invariant. Show that, conversely, the time-ordering for two events with a space-like interval of separation is not necessarily preserved in Lorentz transformations.

7.3   An observer O' in Σ' builds a clock that consists of two mirrors facing one another, each fixed at the end of a metre rule. This observer is able to inject a flash of light between the mirrors so that it travels backwards and forwards between them. Every reflection at mirror 1 generates a tick of this clock. He stations this clock so that the metre rule lies along the x'-axis in Σ'. Observers in Σ will see that the flash of light bouncing between the two mirrors used by O' does a zig-zag course in Σ. Observers in Σ have built clocks identical to that constructed by O' and have synchronised them. A set of these clocks is stationed along the path of mirror 1 in Σ and are able to report later the synchronised times in Σ of the successive ticks of the moving clock. Use this situation and the invariance of the speed of light to derive the formula for time dilation.

7.4 A beam of charged pions, momentum $300\,\text{GeV}/c$, has a free-flight path between two markers that are $600\,\text{m}$ apart in a laboratory. Calculate the speed of the pions in the laboratory and the laboratory time taken to travel the $600\,\text{m}$. Consider observers in the inertial frame in which the pions are at rest. What distance would they find between the markers? What time interval would each measure between the passing of the two markers? Calculate the fraction of the pions that decay in traversing that 600 metres. The mass of the charged pion is $139.6\,\text{MeV}/c^2$ and it has a mean life at rest of $2.603 \times 10^{-8}\,\text{s}$.

7.5 The Lorentz transformation of the time–space coordinates of an event from Σ' to Σ is given by equation (7.5). If $\mathbf{v}'$ is the velocity of a particle in Σ' then the four quantities $(\gamma_{v'}c, \gamma_{v'}\mathbf{v}')$, where $\gamma_{v'} = 1/\sqrt{1 - (v'/c)^2}$, transform to Σ as do the four quantities $(ct', \mathbf{r}')$. Show that this recipe leads to the same results as given in equations (7.17)–(7.19) for the transformation of the components of $\mathbf{v}'$.

7.6 On New Year's day 2030 astronaut Pollux sets out from the Earth at a speed of $0.8c$ to travel to our nearest star α-Centauri, a distance four light years away (as measured in the Earth's inertial frame). On reaching the star, Pollux immediately turns around and returns to Earth, arriving on New Year's Day 2040 (Earth time). Pollux has a twin Castor who remains on the Earth. They have agreed to send each other greetings by radio every New Year's Day until Pollux returns.

   a Satisfy yourself that Pollux sends only six messages (including the one on the last day of the trip), whereas Castor sends 10.

   b Draw a space–time diagram of Pollux's journey in the Earth frame. (Mark the scales ct and distance in years and light-years respectively.) Draw the world lines of all the radio signals that Castor transmits. Use the diagram to verify that Pollux has received only one message up to the moment of reaching the star and, turning back, receives the other nine during the return half of his trip.

   c On the same diagram but using a different colour of pen ink, draw the world lines of all the radio signals that Pollux sends. Verify that Castor receives one message each three years for the first nine years of Pollux's absence, and then three in the last year to make a total of six. This is the expected total since according to Pollux's time the trip takes three years out and three years back. (C. Darwin, *Nature* **180**, 1957, pp. 976–7.)

7.7 This problem is a continuation of problem 6.16 and needs its numerical results. An important decay mode of the Υ meson (mass $10.580\,\text{GeV}/c^2$) is

$$\Upsilon \to B^0 + \overline{B}^0,$$

where B^0 and $\overline{B}^0$ are also mesons.

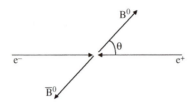

The figure shows the decay of the Υ in the centre-of-mass frame of the colli-sion of an electron and a positron and in which this particle is formed at rest. The outgoing B^0 is moving at an angle θ with respect to the direction of the incident electron. The B^0 and the $\overline{B}^0$ have the same rest mass, $5.279\,\mathrm{GeV}/c^2$. Calculate the value of

1 the momentum of the B^0, and
2 the mean distance moved by the B^0 mesons before decay,

both in the rest frame of the Υ. The mean life of the B^0 is 1.56×10^{-12} s.

Find an expression for the transverse and longitudinal momentum of the B^0 in the laboratory (defined in Problem 6.16) as a function of the angle θ. Show that all the B^0 (and all the $\overline{B}^0$) are produced with their momentum vectors within a cone of half angle less than about $7°$ having its axis along the direction of the incident electrons. Calculate the extreme values of the momentum of the B^0 mesons in the laboratory and hence find the expected range of the mean distances moved by the B^0 mesons in the laboratory.

Comment on the advantage for the measurement of the B^0 lifetime of the asymmetric beam energies compared to a design with equal beam energies.

7.8 a A particle moves with momentum $\mathbf{P}$ at an angle of $45°$ to the z-axis in Σ. What angle does its trajectory make with the z'-axis in Σ'?
 b A rod is stationary in Σ at $45°$ to the z-axis. What angle does it make with the z'-axis in Σ'?

7.9 Two observers A and B, each in an inertial frame, have identical clocks. B passes close to A with a relative speed $0.5c$. A time T seconds after B has passed, A flashes him a light signal (as measured on his own clock). A time $2T$ seconds after receiving this signal (as measured on his own clock) B acknowledges by sending a return light signal to A. When does A receive it?

7.10 Consider three identical clocks, A, B, and C. B is moving relative to A with constant speed u. C is moving in the same direction as B with constant speed v ($v > u$) relative to A. B passes A at which event clocks A and B record the times $t_A = 0 = t_B$. Later C passes A and the clocks A and C record the times $t_A = T_1 = t_C$ for this event. When clock A reads T_1, B has reached a distance d from A measured in A's rest frame.

1 What is the time T_1?
2 What is the time T_2 recorded by clock A when C catches up with B?

3 What is the time T_3 recorded by clock C when it catches up with B? Give all times in terms of u, v, d, and c, the speed of light.

7.11 Two rockets move apart with equal and opposite velocities of magnitude $c/2$ as measured by an observer stationed half way between the rockets. What is the speed of one rocket as measured in an inertial frame in which the other is at rest?

   If each rocket carries an operating radio transmitter of frequency f_0 as measured in its own rest frame, what frequency will be received (a) by the observer and (b) by each rocket.

7.12 Returning to problem 6.5, what wavelength does the pilot of one shuttle see from the headlights of the other as the two shuttles approach one another?

7.13 Show that two successive Lorentz transformations with speed u_1 and u_2 in the same direction are equivalent to a single transformation with speed

$$(u_1 + u_2)/(1 + u_1 u_2/c^2).$$

7.14 A science fiction question: John sees Jane running towards him at a speed of $8\,\text{km}\,\text{h}^{-1}$ and directly away from the Andromeda Galaxy. As she passes him she says "A battle fleet has this moment set out from Andromeda with the intention of destroying the Earth and with it all mankind." John replies "I reckon that battle fleet left Andromeda six days ago!".

   Could they both be correct? (The Andromeda galaxy is approximately 2,200,000 light years from the Earth.)

7.15 A rod has a proper length of 1 m and lies parallel to the z-axis in Σ. It moves non-relativistically with speed v, in a direction at $5°$ to the z-axis (exaggerated in the diagram). It is heading for a slot 1.1 m wide through which it can pass.

Now suppose $v = 0.8\,c$. Does the rod pass through? Describe the reconstruction of what happens as observed

   i from Σ, from which the rod appears contracted, and
   ii from Σ', which is the frame in which the rod is at rest and from which the slot appears contracted.

7.16 The problem is to find the acceleration **a** of a particle in Σ in terms of its acceleration **a′** in Σ'. The accelerations are defined by

$$a_x = \frac{dv_x}{dt} = \frac{dv_x/dt'}{dt/dt'}, \quad a'_{x'} = \frac{dv'_{x'}}{dt'}, \quad \text{and so on.}$$

where the velocity of the particle is $\mathbf{v}$ in Σ and $\mathbf{v}'$ in Σ'. Starting from the results of Section 7.15 show that

$$a_x = \frac{1}{\gamma_u^2(1 + uv'_{z'}/c^2)^2}\left\{a'_x - \frac{uv'_{x'}/c^2}{1 + uv'_{z'}/c^2}a'_z\right\},$$

$$a_y = \frac{1}{\gamma_u^2(1 + uv'_{z'}/c^2)^2}\left\{a'_y - \frac{uv'_{y'}/c^2}{1 + uv'_{z'}/c^2}a'_z\right\},$$

$$a_z = \frac{1}{\gamma_u^3(1 + uv'_{z'}/c^2)^3}a'_z.$$

7.17 Inertial frame Σ' moves with speed u in Σ. Draw the Minkowski map relating Σ' and Σ.

1 Draw the hyperbolae that allow you to define unit distance along the z-, z'-, ct- and ct'-axes.

2 Use the diagram to demonstrate the symmetrical nature of the Lorentz–FitzGerald contraction and of time dilation between Σ' and Σ.

8 The formalities of special relativity

**8.1 Introduction

We trust that the reader is now comfortable with the Lorentz transformations. We have shown that they are consistent with Einstein's postulates. We now derive the transformations from the postulates. We also justify some matters stated as facts in earlier chapters, examine some of the formal properties of these transformations and introduce the important matter of four-vectors.

We shall avoid discussing one matter in deriving the Lorentz transformations. It concerns the minimum number of postulates and assumptions that must be formulated in order to carry through a derivation in the most economical manner. In fact there are several ways of obtaining the transformation equations. We try to be simple and straightforward without leaving any obvious flaws in the logic.

**8.2 The derivation of the Lorentz transformation

We shall derive the equations that connect the coordinates of an event in inertial frame Σ' with its coordinates in inertial frame Σ of our standard arrangement shown in Figure 7.1 (and 5.1). The equations (7.4) for this Lorentz boost along the z-axis that are to be derived are

$$ct = \gamma_u(ct' + \beta_u z'),$$

$$x = x',$$

$$y = y',$$

$$z = \gamma_u(z' + \beta_u ct').$$

As usual $t = t' = 0$ when the origins of the two frames coincide; u is the speed of Σ' in Σ and $\beta_u = u/c$, $\gamma_u = 1/\sqrt{1 - \beta_u^2}$.

These equations follow from Einstein's two postulates (Section 5.5) which we reproduce here:

EP1. All inertial frames are equivalent with respect to the laws of physics.
EP2. The speed of light in empty space is independent of the state of motion of its source.

In order to proceed we have to make some statements of a more specific kind than EP1 and 2. Throughout we assume that the method of synchronisation of clocks is universal to all inertial frames. Our statements are as follows:

S1. Time *flows* in one direction. The only macroscopic processes that indicate a direction for the arrow of time are those governed by the second law of thermodynamics. A simple statement of that law is 'No process is possible whose sole result is the transfer of heat from a cold body to a body that is hotter.' The idea of the flow of heat involves time and so we can label time as increasing if a cold body becomes warmer when it is in contact with one that is hotter. EP1 tells us this will be a valid indicator of the direction of time in all inertial frames.

S2. Proper quantities are invariant. *Proper quantities* (Section 5.1) are those measured by an observer in an inertial frame and in circumstances in which the body exhibiting those quantities is at rest in that frame. The invariance implies that the standards of mass, length and time are the same in all inertial frames. Then it follows, for example, that all observers in different inertial frames will measure the same value for the mean lifetime of a specified excited state of a given atomic species, and the same value for the wavelength of a spectral line that might be emitted on spontaneous de-excitation of this species of atom.

S3. Space (empty) is the same in all directions (isotropic) and at all places (homogeneous). This follows from EP2. Apart from the constancy of the speed of light, this means that a displacement of the origin or a change in the orientation of an inertial reference frame has no effect on any physics described in such a frame. The constancy allows the definition of length to be given by a time interval and the fixed, exact number for the speed of light (Section 5.2).

S4. The transformation equations are linear. That means that z, for example, can only depend on t', x', y', z' and not on terms such as x'^2 or $y'z'$. To see this consider a particle in uniform motion in Σ'. From Einstein's postulate EP1 and the properties of inertial frames it follows that its motion in Σ must also be uniform. That means that a straight line in Σ' must be a straight line in Σ and the coordinates of the spatial position of the particle in Σ' are a linear function of time, and likewise in Σ. The simplest way of satisfying these requirements is by a linear transformation of space–time coordinates between Σ' and Σ.

S5. Reciprocity holds. This means that if the origin of Σ' is moving with speed u along the z-axis of Σ, then an observer in Σ' will find that the origin of Σ is moving with speed u in the direction of the negative z'-axis in Σ'.

S6. The transformation equations must become those (equations (5.2), (5.3) and (7.1)) for a Galilean transformation when $u/c \rightarrow 0$ and the event coordinates remain finite. (The last proviso has to be included so that the term $\beta_u z'$ in the equation for ct does not survive even if u/c becomes very small.) When $u = 0$ the transformation equations must become equalities between corresponding coordinates.

Table 8.1 A summary of the postulates and statements that we use in the derivation of the Lorentz transformations

EP1	All inertial frames are equivalent with respect to the laws of physics.
EP2	The speed of light in empty space is independent of the state of motion of its source.
S1	Time flows in one direction in all inertial frames.
S2	Proper quantities are invariant.
S3	Space is isotropic and homogeneous.
S4	The Lorentz transformations are linear.
S5	Reciprocity holds.
S6	Lorentz becomes Galilean as $u/c \rightarrow 0$ and the event coordinates remain finite.

Note that we have labelled Einstein's postulates (EP1, 2) and the statements (S1–6), and will refer to them by these labels. We summarise them in Table 8.1. The statements may seem to be merely obvious or sensible to the reader and we have already taken some of them for granted when we quoted the equations for the Galilean transformations in Section 5.3. We do not need to examine them in detail.

For the boost along the z-axis of Figure 7.1, the derivation can be divided into two parts:

a That for the transformation of the time and z coordinates.
b That for the transformation of the coordinates (x and y) transverse to the direction of the boost.

We give two derivations for 'part a' (Sections 8.2 and 8.3). 'Part b' is dealt with in Section 8.4. Throughout, S2 expresses an indispensible condition.

As in Section 7.4, since the speed of light in empty space is a universal constant, we will continue to make the algebra easier by having the time coordinate t associated with a c, that is use ct instead of t.

8.3   Derivation using the k-calculus

This is an interesting and simple derivation. However, it is given only in a few books.[1] Readers may wish to go straight to the derivation in Section 8.4 which is one of the more usual.

The k-calculus uses events and world lines in the space–time map for the ct and z coordinates (Section 7.5). Recall that the world line for a photon moving along the positive or negative z-axis lies at $+45°$ or $-45°$ to the ct-axis. The world line for an object that has rest mass always has a tangent that lies at an angle less than $45°$ to the ct-axis. See Figure 7.2.

1 Rosser, W. G. V., *Introductory special relativity*, London: Taylor & Francis, 1991. D'Inverno, R., *Introducing Einstein's relativity*, Oxford: Clarendon Press, 1996. Bohm, D., *The special theory of relativity*, London: Routledge, 1996.

We need to think about two clocks, C stationary at the origin of Σ, and C′ stationary at the origin of Σ′. C is capable of emitting very short flashes of light directed along the positive z-axis, of detecting flashes that are reflected back to it, and of recording the time of such events. C′ is capable of detecting a flash of light that passes it in either direction along the z'-axis, or of reflecting it, and of recording the time of such events.

In using the times from these clocks we shall be relying on S2 and S3 to be sure that these times are in the same units in Σ and in Σ′, and that time intervals may be transformed into distances in a uniform and consistent manner everywhere.

The derivation breaks into three steps.

Step 1. Determining the coordinates of a reflection at a mirror.

Look at Figure 8.1. The world line of C lies along the ct-axis. At event $\mathcal{A}$ (coordinates in Σ: $(ct, z)_{\mathcal{A}} = (cT_{\mathcal{A}}, 0)$) C emits a flash of light which is reflected by the mirror at event $\mathcal{P}$ and returns towards C. It arrives at event $\mathcal{B}$, coordinates $(ct, z)_{\mathcal{B}} = (cT_{\mathcal{B}}, 0)$. This situation will remind the reader of Einstein's

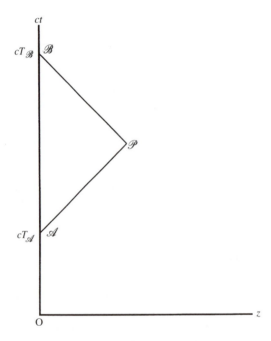

Figure 8.1 The map of the ct–z plane of the inertial frame Σ. The world line of a clock C stationary at $z = 0$ is along the ct-axis. There are three events: at event $\mathcal{A}$ a light flash is emitted from the clock along the z-axis, reflected at event $\mathcal{P}$ by a mirror stationed on the z-axis, and received back at the origin and clock C at event $\mathcal{B}$. The world line of photons in the flash is along $\mathcal{A}\mathcal{P}$ outwards and along $\mathcal{P}\mathcal{B}$ on the return to $\mathcal{B}$. Note we are using script capitals to label events ($\mathcal{A}$, $\mathcal{B}$, …, $\mathcal{P}$, $\mathcal{Q}$, …) and to mark their place on the Minkowski map.

prescription for the synchronisation of clocks, Section 5.7. That means we can deduce the coordinates $(ct, z)_{\mathcal{P}}$ of event $\mathcal{P}$ in Σ in terms of $T_{\mathcal{A}}$ and $T_{\mathcal{B}}$ (S3):

$$\text{In } \Sigma: \quad (ct, z)_{\mathcal{P}} = \left(\tfrac{1}{2}c[T_{\mathcal{B}} + T_{\mathcal{A}}], \tfrac{1}{2}c[T_{\mathcal{B}} - T_{\mathcal{A}}]\right). \tag{8.1}$$

Step 2. Concerning the k-factor.
Look at Figure 8.2. $\mathcal{O}$, $\mathcal{A}$, and $\mathcal{B}$ are events on the world line of clock C in Σ. $\mathcal{O}$ and $\mathcal{P}$ are events on the world line of the clock C'. $\mathcal{AP}$ is the world line of a flash of light which is emitted by C at event $\mathcal{A}$. And $\mathcal{PB}$ is the world line of the flash which is reflected at event $\mathcal{P}$ by the clock C' and returns to the clock C at event $\mathcal{B}$. We can assign specific values to the coordinates of events:

$$\text{In } \Sigma: \quad (ct, z)_{\mathcal{O}} = (0, 0),$$
$$(ct, z)_{\mathcal{A}} = (cT_{\mathcal{A}}, 0),$$
$$(ct, z)_{\mathcal{B}} = (cT_{\mathcal{B}}, 0),$$
$$\text{In } \Sigma': \quad (ct', z')_{\mathcal{O}} = (0, 0),$$
$$(ct', z')_{\mathcal{P}} = (cT'_{\mathcal{P}}, 0).$$

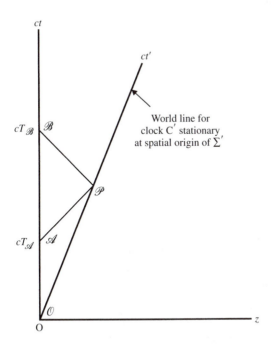

Figure 8.2 This is almost the same as Figure 8.1 except that the event $\mathcal{P}$ is on the world line of a clock C' stationary at the spatial origin of inertial frame Σ' that is moving with speed u along the z-axis of Σ.

Consider now the time interval from O to $\mathcal{P}$, $T'_{\mathcal{P}}$, measured on the clock C', and the time interval $T_{\mathcal{A}}$ from O to $\mathcal{A}$ measured on clock C. The relation between $T_{\mathcal{A}}$ and $T'_{\mathcal{P}}$ (a) can only be linear (S4); (b) cannot depend explicitly on the z coordinates of events $\mathcal{A}$ and $\mathcal{P}$ (S3) and (c) must satisfy $T'_{\mathcal{P}} = 0$ when $T_{\mathcal{A}} = 0$.

Therefore, the only possibility is

$$T'_{\mathcal{P}} = k\, T_{\mathcal{A}}, \tag{8.2}$$

where k is independent of the times involved. This is the k-factor. And it must be >0.

The k-factor must apply in both directions between the frames Σ and Σ' (postulate EP1) therefore the time interval from O to $\mathcal{B}$, $T_{\mathcal{B}}$, stands to the interval from O to $\mathcal{P}$, $T'_{\mathcal{P}}$, as does the latter interval to that from O to $\mathcal{A}$, $T_{\mathcal{A}}$, therefore we have

$$cT_{\mathcal{B}} = k\, c T'_{\mathcal{P}} = k^2\, c T_{\mathcal{A}}. \tag{8.3}$$

We now use equation (8.1) to find the coordinates in Σ of the event $\mathcal{P}$:

$$\text{In } \Sigma: \quad (ct, z)_{\mathcal{P}} = \left(\tfrac{1}{2}c\left[k^2\, T_{\mathcal{A}} + T_{\mathcal{A}}\right], \tfrac{1}{2}c\left[k^2\, T_{\mathcal{A}} - T_{\mathcal{A}}\right] \right).$$

But $\mathcal{P}$ is on the world line of clock C' moving with speed u in Σ so that for this event

$$z = ut$$

or

$$\frac{z}{ct} = \frac{u}{c} = \frac{(k^2 - 1)}{(k^2 + 1)}.$$

Hence

$$k = \sqrt{\frac{1 + u/c}{1 - u/c}}. \tag{8.4}$$

This is equivalent to the formula for the longitudinal Doppler effect in which the observed time period of the oscillation of a light source is increased when it is receding from the observer. In the present situation the clock C appears to be receding from C'. Two light flashes emitted by C at O and at $\mathcal{A}$, with time interval $T_{\mathcal{A}}$, are observed by C' with a time interval $T'_{\mathcal{P}} = kT_{\mathcal{A}} > T_{\mathcal{A}}$.

Step 3. A result from the k-calculus.

Look at the events depicted in Figure 8.3. As before the events O, $\mathcal{A}$ and $\mathcal{B}$ lie on the world line of clock C. Similarly, events O, $\mathcal{L}$ and $\mathcal{M}$ lie on the world line of clock C'. A flash of light emitted by C at event $\mathcal{A}$ overtakes C' at event $\mathcal{L}$ and is reflected back at event $\mathcal{2}$. It meets and passes clock C' at event $\mathcal{M}$ and returns to clock C at event $\mathcal{B}$. So $\mathcal{ALQ}$ is the world line for the emitted light flash and $\mathcal{2M\mathcal{B}}$

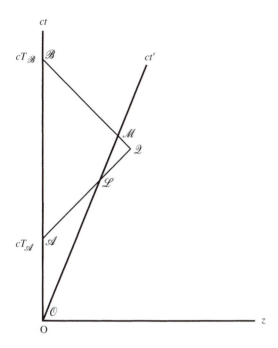

Figure 8.3 The light flash from event $\mathcal{A}$ passes the clock C′ at event $\mathcal{L}$, is reflected by a mirror at event $\mathcal{D}$. At event $\mathcal{M}$ the light flash meets the clock C′ during its return journey to clock C at event $\mathcal{B}$.

is the world line for the reflected flash. Now we write down some specific event coordinate values:

In Σ: $(ct, z)_{\mathcal{O}} = (0, 0)$,

$(ct, z)_{\mathcal{A}} = (cT_{\mathcal{A}}, 0)$,

$(ct, z)_{\mathcal{B}} = (cT_{\mathcal{B}}, 0)$,

In Σ': $(ct', z')_{\mathcal{O}} = (0, 0)$.

Then it follows as in step 2 (equations (8.2) and (8.3)) that

In Σ': $(ct', z')_{\mathcal{L}} = (kcT_{\mathcal{A}}, 0)$,

$(ct', z')_{\mathcal{M}} = (cT_{\mathcal{B}}/k, 0)$.

We now use equation (8.1) to find the coordinates of event $\mathcal{D}$ in Σ in terms of $T_{\mathcal{A}}$ and $T_{\mathcal{B}}$:

In Σ: $(ct, z)_{\mathcal{D}} = \left(\frac{1}{2}c[T_{\mathcal{B}} + T_{\mathcal{A}}], \frac{1}{2}c[T_{\mathcal{B}} - T_{\mathcal{A}}]\right).$ \hfill (8.5)

The same procedure is now used in Σ' to find the coordinates of event $\mathcal{D}$ in that frame in terms of the times (again in Σ) of events $\mathcal{L}$ and $\mathcal{M}$. We can do that because equation (8.1) is true in any inertial frame (EP1) and because the events $\mathcal{L}$, $\mathcal{M}$, and $\mathcal{Q}$ are in the same relation to one another in Σ' as the events $\mathcal{A}$, $\mathcal{B}$ and $\mathcal{P}$ of Figure 8.1 are in Σ. Thus

$$\text{In } \Sigma': \quad (ct', z')_\mathcal{Q} = \left(\tfrac{1}{2}c[T_\mathcal{B}/k + kT_\mathcal{A}], \tfrac{1}{2}c[T_\mathcal{B}/k - kT_\mathcal{A}]\right). \tag{8.6}$$

Equation (8.5) gives, for $cT_\mathcal{B}$ and $cT_\mathcal{A}$ in terms of $(ct, z)_\mathcal{L}$

$$cT_\mathcal{B} = ct + z \quad \text{and} \quad cT_\mathcal{A} = ct - z.$$

Substituting for $T_\mathcal{A}$, $T_\mathcal{B}$ and k in equation (8.6) gives for the coordinates of event $\mathcal{Q}$:

$$ct' = \frac{1}{\sqrt{1 - u^2/c^2}}(ct - uz/c) = \gamma_u(ct - \beta_u z) \tag{8.7}$$

and

$$z' = \frac{1}{\sqrt{1 - u^2/c^2}}(z - ut) = \gamma_u(z - \beta_u ct), \tag{8.8}$$

where, as usual,

$$\gamma_u = 1/\sqrt{1 - u^2/c^2} \quad \text{and} \quad \beta_u = u/c.$$

These equations can be written for the inverse transformation by interchanging primed to unprimed coordinates and replacing $-u$ or $-\beta_u$ with $+u$ or $+\beta_u$. In matrix form that transformation is:

$$\begin{pmatrix} ct \\ z \end{pmatrix} = \begin{pmatrix} \gamma_u & \gamma_u \beta_u \\ \gamma_u \beta_u & \gamma_u \end{pmatrix} \begin{pmatrix} ct' \\ z' \end{pmatrix}. \tag{8.9}$$

The Lorentz factor $\gamma_u \rightarrow 1$ as $u/c \rightarrow 0$. Therefore these transformations become the Galilean transformations (S6) when $u \ll c$ and $z'u/c^2$ (or zu/c^2) $\rightarrow$ 0, as discussed in Section 7.2.

This completes the k-calculus derivation of half what was quoted in equations (7.4).

**8.4   Derivation using a conventional method

As before we have to find the relation between the coordinates (ct, z) in the frame Σ of an event and the coordinates (ct', z') in the frame Σ' of the same event. Since the relation is linear (S4) and since the origins of these coordinate systems coincide when $t = 0 = t'$, we can write the relation in a simple matrix form:

$$\begin{pmatrix} ct \\ z \end{pmatrix} = \begin{pmatrix} A & B \\ C & D \end{pmatrix} \begin{pmatrix} ct' \\ z' \end{pmatrix}.$$

The reverse transformation will be effected by the inverse of the square matrix:

$$\begin{pmatrix} ct' \\ z' \end{pmatrix} = \frac{1}{AD - BC} \begin{pmatrix} D & -B \\ -C & A \end{pmatrix} \begin{pmatrix} ct \\ z \end{pmatrix}.$$

We expect the quantities A, B, C, D to be functions of the relative speed u.

i   Consider the origin of Σ' which has $z' = 0$ at all t'. In Σ it moves with speed u along the z-axis and has coordinates in that frame (ct, z) related by the equation $z = ut$. Therefore

$$z' = \frac{Az - Cct}{AD - BC} = \frac{(Au - Cc)t}{AD - BC}.$$

But $z' = 0$ for the origin of Σ' and it follows that

$$C/A = u/c. \tag{8.10}$$

ii   Consider the origin of Σ which has $z = 0$. In Σ' it moves with speed $-u$ (S5) along the z'-axis and has coordinates in that frame (ct', z') related by the equation $z' = -ut'$. Therefore

$$z = Cct' + Dz' = Cct - Dut.$$

But $z = 0$ for the origin of Σ and it follows that

$$C/D = u/c. \tag{8.11}$$

iii   From equations (8.10) and (8.11) we conclude that

$$D = A. \tag{8.12}$$

iv   Event $\mathcal{O}$ occurs when the origins of Σ and Σ' coincide. In Σ this event has coordinates $(ct, z)_{\mathcal{O}} = (0, 0)$, and in Σ' it has coordinates $(ct', z')_{\mathcal{O}} = (0, 0)$. Suppose that at event $\mathcal{O}$ a flash of light is emitted along the positive $z(z')$-axis. Later the light flash has reached an event $\mathcal{P}$. In Σ this event has coordinates $(ct, z)_{\mathcal{P}}$ where $z = ct$ and in Σ' the coordinates are $(ct', z')_{\mathcal{P}}$ where $z' = ct'$.

Now

$$ct = Act' + Bz' = (A + B)ct'$$

and

$$z = Cct' + Dz' = (C + D)ct'.$$

But $z = ct$ and $D = A$ (equation (8.12)) therefore

$$B = C \tag{8.13}$$

and from equation (8.10)

$$B/A = u/c. \tag{8.14}$$

v   The two transforming equations now appear as

$$\begin{pmatrix} ct \\ z \end{pmatrix} = \begin{pmatrix} A & B \\ B & A \end{pmatrix} \begin{pmatrix} ct' \\ z' \end{pmatrix} \tag{8.15}$$

and

$$\begin{pmatrix} ct' \\ z' \end{pmatrix} = \frac{1}{A^2 - B^2} \begin{pmatrix} A & -B \\ -B & A \end{pmatrix} \begin{pmatrix} ct \\ z \end{pmatrix} \tag{8.16}$$

with

$$B/A = u/c.$$

vi   These two transformations must change one into the other under the simultaneous changes $u \leftrightarrow -u$, $z \leftrightarrow z'$, and $t \leftrightarrow t'$. Consider the change from equation (8.15) to (8.16). In the square matrix, A must be positive: otherwise, increasing t (or t') would lead to decreasing t' (or t) (S1). Thus the 2×2 matrix in equation (8.15) changes thus:

$$\begin{pmatrix} ct' \\ z' \end{pmatrix} = \begin{pmatrix} A & -B \\ -B & A \end{pmatrix} \begin{pmatrix} ct \\ z \end{pmatrix}.$$

Compare this with equation (8.16) and we see that

$$\frac{1}{A^2 - B^2} \begin{pmatrix} A & -B \\ -B & A \end{pmatrix} = \begin{pmatrix} A & -B \\ -B & A \end{pmatrix}.$$

Therefore

$$A^2 - B^2 = 1$$

and

$$A = 1/\sqrt{1 - B^2/A^2} = 1/\sqrt{1 - u^2/c^2} = \gamma_u,$$
$$B = uA/c = \gamma_u \beta_u.$$

vii   Summarising

$$\begin{pmatrix} ct \\ z \end{pmatrix} = \begin{pmatrix} \gamma_u & \gamma_u \beta_u \\ \gamma_u \beta_u & \gamma_u \end{pmatrix} \begin{pmatrix} ct' \\ z' \end{pmatrix} \tag{8.17}$$

and with $\beta_u \rightarrow -\beta_u$ in the matrix for the inverse transformation.

These transformations satisfy S6. Therefore we have derived two of equations (7.4).

**8.5   Derivation of the transformation of the transverse coordinates

Transverse means perpendicular to the direction of the Lorentz transformation. Therefore the coordinates referred to are x and y in Figure 7.1. We consider some events occurring in that standard arrangement of inertial reference frames. In Σ' a clock C' is stationary at the origin and is capable of emitting short flashes of light and of recording the time of emission of a flash and the time of return if the flash is reflected back to the origin. The origins of the two reference frames coincide at $t' = 0 = t$. We call that event $\mathbb{O}$ and at that event C' emits a flash of light along the direction of the y'-axis. This flash is reflected back to C' by stationary mirror on the y'-axis (see Figure 8.4(a)). The flash is received by C' at event $\mathscr{P}$ when $t' = T'$. Summarising we have for the coordinates (ct', z') in Σ' for these events:

In Σ': $(ct', z')_{\mathbb{O}} = (0, 0),$

$(ct', z')_{\mathscr{P}} = (cT', 0).$

The observer in Σ' will deduce that the mirror is at a distance $cT'/2 \ (\equiv Y')$ from the origin along the y'-axis. In Σ the observer will compile the coordinates of these events and reconstruct the path of the light as in Figure 8.4(b). We use equation (8.17) to find these coordinates in terms of those in Σ':

In Σ: $(ct, z)_{\mathbb{O}} = (0, 0),$

$(ct, z)_{\mathscr{P}} = (cT, Z) = (\gamma_u cT', \ \gamma_u \beta_u cT').$

The observer in Σ has to calculate the y coordinate of the moving mirror at the instant it reflects the light flash. Call it Y. Figure 8.4(b) gives the total distance d travelled by the light in Σ:

$$d = 2\sqrt{Y^2 + (Z/2)^2} = 2\sqrt{Y^2 + (\gamma_u \beta_u cT'/2)^2}.$$

The time taken to traverse that distance is T given by

$$cT = \gamma_u cT'.$$

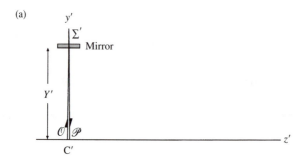

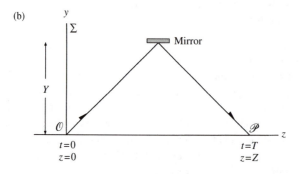

Figure 8.4 (a) A light flash is emitted (event $\mathcal{O}$) from the clock C' at time $t' = 0$ at the origin of Σ', travels along the y'-axis, is reflected at the mirror and returns to the clock C' at time $t' = T'$ (event $\mathcal{P}$). (b) The same events as observed from frame Σ in which Σ' is moving with speed u. The origins of Σ and Σ' coincide at $t' = 0 = t$ (event $\mathcal{O}$).

The speed of light is c to the Σ observer so it must be that $cT = d$. Substituting $d = \gamma_u cT'$ gives:

$$(\gamma_u cT')^2 = 4[Y^2 + (\gamma_u \beta_u cT'/2)^2].$$

Solving we find

$$Y = cT'/2 = Y'.$$

So for the general coordinate we must have

$$y = y' \qquad (8.18)$$

and similarly

$$x = x'. \qquad (8.19)$$

Therefore the transverse spatial coordinates of an event are invariant.

*8.6 Summarising the Lorentz transformations

First, we tie together all the bits and pieces of our derivation (equations (8.9) or (8.17)–(8.19)) and we have:

$$
\begin{pmatrix} ct \\ x \\ y \\ z \end{pmatrix} = \begin{pmatrix} \gamma_u & 0 & 0 & \gamma_u \beta_u \\ 0 & 1 & 0 & 0 \\ 0 & 0 & 1 & 0 \\ \gamma_u \beta_u & 0 & 0 & \gamma_u \end{pmatrix} \begin{pmatrix} ct' \\ x' \\ y' \\ z' \end{pmatrix} = \mathbf{L}(\beta_u) \begin{pmatrix} ct' \\ x' \\ y' \\ z' \end{pmatrix}, \tag{8.20}
$$

where u is the speed of Σ' in Σ as in our standard arrangement, $\gamma_u = 1/\sqrt{1 - \beta_u^2}$, $\beta_u = u/c$. For the inverse transformation change $\beta_u \rightarrow -\beta_u$ and interchange primed and unprimed coordinates. Equation (8.20) is the same as equation (7.5). In Sections 7.2 and 7.3 we discussed some of the properties of these transformations.

Our derivation of the Lorentz transformation is for a boost. A boost is a transformation from one inertial reference frame Σ to a second Σ' with:

1 A relative velocity $\mathbf{u}$ of Σ' in Σ.
2 An event for which the spatial origins and axes of Σ and Σ' coincide.
3 That particular event is usually defined to have $t = 0 = t'$ in the case of a boost of the coordinates of an event.

Thus, our derivation is for a boost along the $z(z')$-axis with speed u. In the next Section, 8.7, we look briefly at the boost with velocity u not necessarily along the $z(z')$-axis.

The inhomogeneous Lorentz transformations have a displacement of the origin in one or more of the time and three space coordinates. We have lost no important physics in deriving a homogeneous transformation by having the spatial Cartesian axes coincide at $t = 0 = t'$.

Both the homogeneous and inhomogeneous transformations can be subdivided into proper and improper transformations. The latter have a space inversion of one or three axes (for example, $x \rightarrow x' = -x$) or a time reversal, $t \rightarrow t' = -t$, or both. These transformations are important for the understanding of the symmetry properties of atomic and sub-atomic systems. That subject would take us beyond the scope of this book; interested readers are referred for an introduction to Martin and Shaw.[2] All our transformations are proper.

The proper transformations have a subset that are rotations in ordinary three-dimensional space. These are straightforward and have no dependence on relativistic principles. Thus a catalogue of Galilean transformations includes such rotations and inhomogeneous transformations.

The proper homogeneous Lorentz transformations can be factored into rotations and boosts.

2 Martin, B. R. and Shaw, G., *Particle physics*, Chichester: John Wiley & Sons, 1992, Chapter 4.

8.7   The general Lorentz boost

Figure 8.5 shows a variation of our standard arrangement of reference frames. Now Σ' is moving in Σ with a velocity $\mathbf{u} = \boldsymbol{\beta}_u c$ instead of speed u parallel to the z-axis in Σ. In both cases there is one event where the origins coincide and the clocks are set to $t = 0 = t'$.

We can find the matrix that transforms coordinates for the situation of Figure 8.5 by starting with the transformation equations for the standard arrangement (Figure 7.1), equations (7.4):

$$ct = \gamma_u(ct' + \beta_u z'),$$

$$x = x',$$

$$y = y',$$

$$z = \gamma_u(z' + \beta_u ct'),$$

where as usual $\beta_u = u/c$ and $\gamma_u = 1/\sqrt{1 - u^2/c^2}$. (We shall, in this and the next sections, use analogous relations for other speeds v and w.) We rewrite these equations in a different form using vectors (with their components) $\mathbf{r} = (x, y, z)$, $\mathbf{r}' = (x', y', z')$, and $\boldsymbol{\beta}_u c = (0, 0, u)$:

$$ct = \gamma_u(ct' + \boldsymbol{\beta}_u \cdot \mathbf{r}'), \tag{8.21}$$

$$\mathbf{r} = \mathbf{r}' + (\gamma_u - 1)(\boldsymbol{\beta}_u \cdot \mathbf{r}')\boldsymbol{\beta}_u/\beta_u^2 + \gamma_u \boldsymbol{\beta}_u ct'. \tag{8.22}$$

Problem 8.2 asks you to check this rewrite.

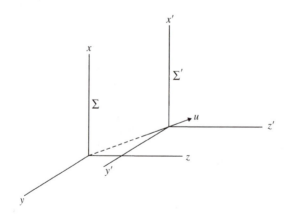

Figure 8.5 A general Lorentz boost. The difference between this and the standard arrangement is that the origin of Σ' is moving with velocity $\mathbf{u} = \boldsymbol{\beta}_u c$ in Σ whereas in the latter the origin of Σ' is moving with speed u along the z-axis of Σ. In both cases there is an event at which the origins coincide and at which, by convention, the clocks at the origins are set to zero. Thus this event has $t = 0 = t'$.

We now change to a general $\boldsymbol{\beta}_u c = \mathbf{u}$ (components u_x, u_y, u_z). Then these last equations give the correct transformation. They look complicated but what they are doing is sorting out the components of the coordinates of an event in Σ' into those parts that are transverse to $\boldsymbol{\beta}_u$ and therefore invariant (as for x and y in equation (8.20)) and others that change (ct and z in equation (8.20)).

This transformation of coordinates can be put into matrix form. We drop the subscript u and just use $\boldsymbol{\beta}$ and γ to avoid a proliferation of subscripts. Thus $\boldsymbol{\beta}$ has components $\beta_x, \beta y$ and β_z in Σ. And $\beta = |\boldsymbol{\beta}|$. The matrix that replaces $\mathbf{L}(\beta_u)$ in equations (6.5), (7.5), and (8.20) for this general boost is $\mathbf{L}(\beta)$ given by:

$$
\mathbf{L}(\boldsymbol{\beta}) = \begin{pmatrix} \gamma & \gamma\beta_x & \gamma\beta_y & \gamma\beta_z \\ \gamma\beta_x & 1 + (\gamma - 1)\beta_x\beta_x/\beta^2 & (\gamma - 1)\beta_x\beta_y/\beta^2 & (\gamma - 1)\beta_x\beta_z/\beta^2 \\ \gamma\beta_y & (\gamma - 1)\beta_y\beta_x/\beta^2 & 1 + (\gamma - 1)\beta_y\beta_y/\beta^2 & (\gamma - 1)\beta_y\beta_z/\beta^2 \\ \gamma\beta_z & (\gamma - 1)\beta_z\beta_x/\beta^2 & (\gamma - 1)\beta_z\beta_y/\beta^2 & 1 + (\gamma - 1)\beta_z\beta_z/\beta^2 \end{pmatrix}.
$$

(8.23)

This matrix has all the properties required to satisfy the postulates of special relativity, just as we found for the equations (7.4) that gave the matrix for the transformation between Σ and Σ' for the standard arrangement of reference frames (equation (7.5)).

The striking thing about $\mathbf{L}(\beta)$ is that it is a symmetric matrix. This is true for an **L** that transforms four-vectors in a boost. However, the symmetry property does not extend to a matrix **L** that transforms four-vectors when a rotation of a reference frame is involved.

+8.8 Two successive Lorentz boosts I

Consider three inertial reference frames Σ, Σ', and Σ''. Σ' is moving with speed u along the z-axis of Σ (our usual arrangement) and Σ'' is moving with speed v along the z-axis in Σ. The transformation of coordinates from Σ' to Σ is effected by the matrix $\mathbf{L}(\beta_u)$, equation (8.19). The transformation from Σ to Σ'' will be effected by the matrix $\mathbf{L}(-\beta_v)$. Now if Σ' is moving with speed w along the z''-axis in Σ'' the transformation of coordinates from Σ' to Σ'' will be effected by the matrix $\mathbf{L}(\beta_w)$. We expect

$$\mathbf{L}(\beta_w) = \mathbf{L}(-\beta_v)\mathbf{L}(\beta_u).$$

Substituting in the right-hand side of this expression for the matrices **L** (equation (8.20)) we find:

$$
\mathbf{L}(\beta_w) = \begin{pmatrix} \gamma_u\gamma_v(1 - \beta_u\beta_v) & 0 & 0 & \gamma_u\gamma_v(\beta_u - \beta_v) \\ 0 & 1 & 0 & 0 \\ 0 & 0 & 1 & 0 \\ \gamma_u\gamma_v(\beta_u - \beta_v) & 0 & 0 & \gamma_u\gamma_v(1 - \beta_u\beta_v) \end{pmatrix}.
$$

(8.24)

Now the matrix $\mathbf{L}(\beta_w)$ is a function of β_w defined by

$$\mathbf{L}(\beta_w) = \begin{pmatrix} \gamma_w & 0 & 0 & \gamma_w\beta_w \\ 0 & 1 & 0 & 0 \\ 0 & 0 & 1 & 0 \\ \gamma_w\beta_w & 0 & 0 & \gamma_w \end{pmatrix}. \tag{8.25}$$

These two equalities (equations (8.24) and (8.25)) are satisfied if

$$\beta_w = \frac{\beta_u - \beta_v}{1 - \beta_u\beta_v} \quad \text{and} \quad \gamma_w \equiv 1/\sqrt{1 - \beta_w^2} = \gamma_u\gamma_v(1 - \beta_u\beta_v). \tag{8.26}$$

The equation for β_w is similar to equation (7.22) which connects the z components of the velocity that is $\mathbf{v}'$ in Σ' and $\mathbf{v}$ in Σ. That is $\mathbf{v}'$ is the relativistic subtraction of speed u along the z-axis from $\mathbf{v}$. Here speed w is for the relativistic subtraction of speed v from speed u. Thus two Lorentz boosts in the sense described at the beginning of this section, with speeds u and v, are the same as one with speed w if the last is the relativistic difference of u and v. A few moments' thought will show that this must be the case given the way the addition of velocities is defined and derived. Look at Section 7.15.

Equation (8.24) has the property that it remains the same if the boosts u and v are applied in the reverse order. That is

$$\mathbf{L}(-\beta_v)\mathbf{L}(\beta_u) = \mathbf{L}(\beta_u)\mathbf{L}(-\beta_v)$$

or

$$\mathbf{L}(-\beta_v)\mathbf{L}(\beta_u) - \mathbf{L}(\beta_u)\mathbf{L}(-\beta_v) = \mathbf{0} \tag{8.27}$$

where $\mathbf{0}$ represents a 4×4 matrix that has all elements equal to zero. In words, equation (8.27) says the matrices $\mathbf{L}(-\beta_v)$ and $\mathbf{L}(\beta_u)$ 'commute'. This is true of all matrices that transform coordinates for boosts that are all in the same direction. However, in general the matrices associated with two proper Lorentz transformations do not commute. This we shall discuss in the next section.

+8.9   Two successive Lorentz boosts II

We consider two successive boosts as shown diagrammatically in Figure 8.6. The first corresponds to our standard arrangement and the transformation of coordinates from Σ to Σ' is done by the matrix $\mathbf{L}(-\boldsymbol{\beta}_u)$. The second boost is along the y'-axis in Σ' with speed v and takes us to Σ''. The transformation of coordinates from Σ' to Σ'' is done by the matrix $\mathbf{L}(-\boldsymbol{\beta}_v)$ where $\boldsymbol{\beta}_v c$ is the speed of Σ'' along the y'-axis in Σ'. (To construct this matrix, use equation (8.23) or shuffle the elements of equation (8.20) in an obvious way.) The transformation of coordinates from

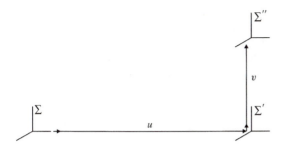

Figure 8.6 A schematic representation of the two successive Lorentz boosts with velocities **u** and **v**, that are discussed in Section 8.9.

Σ to Σ'' is done by the matrix, that is the product, given by

$$
\mathbf{L}(-\boldsymbol{\beta}_v)\mathbf{L}(-\boldsymbol{\beta}_u) =
\begin{pmatrix}
\gamma_v & 0 & -\gamma_v\beta_v & 0 \\
0 & 1 & 0 & 0 \\
-\gamma_v\beta_v & 0 & \gamma_v & 0 \\
0 & 0 & 0 & 1
\end{pmatrix}
\begin{pmatrix}
\gamma_u & 0 & 0 & -\gamma_u\beta_u \\
0 & 1 & 0 & 0 \\
0 & 0 & 1 & 0 \\
-\gamma_u\beta_u & 0 & 0 & \gamma_u
\end{pmatrix}
$$

$$
=
\begin{pmatrix}
\gamma_u\gamma_v & 0 & -\gamma_v\beta_v & -\gamma_u\beta_u\gamma_v \\
0 & 1 & 0 & 0 \\
-\gamma_u\gamma_v\beta_v & 0 & \gamma_v & \gamma_u\beta_u\gamma_v\beta_v \\
-\gamma_u\beta_u & 0 & 0 & \gamma_u
\end{pmatrix}. \tag{8.28}
$$

If the boosts had been performed in the reverse order the corresponding matrix is

$$
\mathbf{L}(-\boldsymbol{\beta}_u)\mathbf{L}(-\boldsymbol{\beta}_v) =
\begin{pmatrix}
\gamma_u\gamma_v & 0 & -\gamma_u\gamma_v\beta_v & -\gamma_u\beta_u \\
0 & 1 & 0 & 0 \\
-\gamma_v\beta_v & 0 & \gamma_v & 0 \\
-\gamma_u\beta_u\gamma_v & 0 & \gamma_u\beta_u\gamma_v\beta_v & \gamma_u
\end{pmatrix}. \tag{8.29}
$$

What do we conclude from equations (8.28) and (8.29)?

1   $\mathbf{L}(-\boldsymbol{\beta}_v)\mathbf{L}(-\boldsymbol{\beta}_u) \neq \mathbf{L}(-\boldsymbol{\beta}_u)\mathbf{L}(-\boldsymbol{\beta}_v)$. That is, these two matrices do not commute.
2   Neither the matrix $\mathbf{L}(-\boldsymbol{\beta}_u)\mathbf{L}(-\boldsymbol{\beta}_v)$ nor the matrix $\mathbf{L}(-\boldsymbol{\beta}_v)\mathbf{L}(-\boldsymbol{\beta}_u)$ is symmetric and therefore neither can be replaced by the transforming matrix required by a single boost. (We know from the last section, Section 8.7, that the latter have symmetric matrices.)

What is the physical significance of these results? To answer this question we have to improve our notation. Until now our standard arrangement has made it easy for us in the sense that Σ and Σ' were frames of reference in different inertial systems. Thus Σ and Σ' could stand for the inertial systems or for the frames of reference. In fact we called them inertial frames thus blurring the distinction. We now need a notation to distinguish reference frames (coordinate systems) that are not the same but are in the same inertial system, that is they have no relative velocity. They will, however, be related by a rotation or a displacement. Here we shall be concerned solely with connection by rotation.

Let us represent two successive Lorentz transformations involving three reference frames as follows:

$$\Sigma \xrightarrow{\mathbf{u}} \Sigma' \xrightarrow{\mathbf{v}} \Sigma''.$$

This clearly means boosts with velocity $\mathbf{u}$ followed by velocity $\mathbf{v}$ which take us from Σ to Σ' to Σ''. The boosts in the reverse order give

$$\Sigma \xrightarrow{\mathbf{v}} S' \xrightarrow{\mathbf{u}} S'',$$

where S' and S'' are not the same reference frames as Σ' and Σ''.

We can shed light on the physical significance by examining the apparent motion of the clock stationed at the origin of Σ as seen by two observers, the first in Σ'' and the second in S''. To do this, transform that clock's Σ coordinates $(ct, \mathbf{r}) = (ct, \mathbf{0})$ into Σ'' and S''.

$$\text{In } \Sigma'': \quad \begin{pmatrix} ct_1 \\ x_1 \\ y_1 \\ z_1 \end{pmatrix} = \mathbf{L}(-\boldsymbol{\beta}_v)\mathbf{L}(-\boldsymbol{\beta}_u) \begin{pmatrix} ct \\ 0 \\ 0 \\ 0 \end{pmatrix} = \begin{pmatrix} \gamma_u \gamma_v ct \\ 0 \\ -\gamma_u \gamma_v \beta_v ct \\ -\gamma_u \beta_u ct \end{pmatrix} \tag{8.30}$$

and

$$\text{In } S'': \quad \begin{pmatrix} ct_2 \\ x_2 \\ y_2 \\ z_2 \end{pmatrix} = \mathbf{L}(-\boldsymbol{\beta}_u)\mathbf{L}(-\boldsymbol{\beta}_v) \begin{pmatrix} ct \\ 0 \\ 0 \\ 0 \end{pmatrix} = \begin{pmatrix} \gamma_u \gamma_v ct \\ 0 \\ -\gamma_v \beta_v ct \\ -\gamma_u \beta_u \gamma_v ct \end{pmatrix}. \tag{8.31}$$

Examining equations (8.30) and (8.31) we can conclude:

1   Since $ct_1 = ct_2$, the time dilation factor between Σ and Σ'' is the same as that between Σ and S''. Therefore the speed of Σ'' in Σ is the same as that of S'' in Σ. That speed is given by

$$\beta c = c\sqrt{1 - (\gamma_u \gamma_v)^{-2}}. \tag{8.32}$$

2   To the observers in Σ'' and S'' the origin O of Σ must be moving with this same speed βc but the components are different:

$$\beta_x c = x_1/t_1 = 0,$$

In Σ'': $\beta_y c = y_1/t_1 = -\beta_v c.$ (8.33)

$$\beta_z c = z_1/t_1 = -\beta_u c/\gamma_v.$$

$$\beta_x c = x_2/t_2 = 0,$$

In S'': $\beta_y c = y_2/t_2 = -\beta_v c/\gamma_u.$ (8.34)

$$\beta_z c = z_2/t_2 = -\beta_u c.$$

3   Therefore equations (8.30) and (8.31) give the equations of motion, in reference frames Σ'' and S'' respectively, of the origin of Σ in terms of the time indicated by the clock stationed at that origin.

4   Equations (8.33) and (8.34) shows us that the velocity $\boldsymbol{\beta}$ has components in Σ'' different from those in S''. Thus Σ'' and S'' are connected by a rotation, in this case, around the x-axis. That axis happens to be common to both frames.

(We have been economical with subscripts. For example, the y-axis implied in β_y of equation (8.33) is not the same y-axis as implied by β_y in equation (8.34). This should cause no difficulties but if in doubt take the non-relativistic limit. There Σ'' becomes the same as S'' and the equations should reflect that fact.)

The reader might ask, 'Why is all this important?' Of course, it is not at an elementary level, but it is useful to know that Lorentz transformations do not in general commute and the consequence can be changes in the apparent direction of vectors after transformations. This property was known early in the development of the theory of special relativity but it became important in 1926 when L. H. Thomas was able to use this property to explain some discrepancies between the observations and the theory of atomic spectra. That episode is mentioned qualitatively in most undergraduate treatments of the subject of atomic spectra. We shall give a simple presentation of the relativity involved in Section 8.13.

Since 1945 the experimental investigation of the physics of elementary particles has been continued at ever increasing particle energies and at speeds increasingly close to c. As stressed in Chapter 3, relativistic kinematics must be used in analysing the behaviour of particles in collisions and decays. In many collisions now studied there can be a large number of particles produced most of which will suffer decay to yet other particles. The analysis of such events may involve transformations between a variety of inertial frames. Under these circumstances it is important to bear in mind that reference frames may not be oriented quite as expected. This is particularly important when the particles have spin, that is, intrinsic angular momentum. The direction in which that angular momentum points in the particle's rest frame does not always behave as might be expected non-relativistically. That shows up in Thomas's solution of the spectroscopy problem (Section 8.13).

The kinds of analyses mentioned in the last paragraph are now performed on large computers. Lorentz transformations may be done by a call for an existing programme sub-routine. It is essential to understand to which reference frame the procedure sends your data!

We can summarise the results of this section by two simple statements:

1   The result of two successive Lorentz transformations may depend on the order in which they are performed.
2   In general, the result of two successive Lorentz transformations is equivalent to a single Lorentz transformation and a rotation.

*8.10   Four-vectors

In Section 7.4 we moved towards combining the time–space coordinates of an event into one object, represented there by a column matrix. In Section 7.15 we introduced the four-velocity which has four components transforming in the same way as the four components of the time–space coordinates $(ct, \mathbf{r})$. These objects are **four-vectors** and in a wider context are rank one contravariant tensors. The reader must not be alienated by this name dropping. The mathematics of tensors is particularly useful in describing Relativity, both Special and General. Therefore, in using four-vectors we feel constrained to use the notation of tensors in order to give the reader some confidence that what may look to be arbitrary definitions have a proper foundation. We shall not trespass very far into tensor land. (See, however, Sections 9.11 and 9.12.)

We consider an event $\mathcal{P}$ in frame Σ of Figure 7.1 which has coordinates ct, x, y, z.

1   The four-vector x that represents the coordinates of $\mathcal{P}$, has four components x^μ, $\mu = 0, 1, 2, 3$, that are, in conventional coordinates,

$$x^0 = ct, \quad x^1 = x, \quad x^2 = y, \quad x^3 = z. \tag{8.35}$$

Be careful! The superscript is not an exponent. We may write this four-vector $x = (ct, \mathbf{r})$.
2   The defining property of any contravariant four-vector is that it Lorentz transforms as does the four-vector x.
3   For three-vectors the scalar product (sometimes called the inner product) of a vector ($\mathbf{r}$ say) with itself is defined by

$$\mathbf{r} \cdot \mathbf{r} = (x)^2 + (y)^2 + (z)^2.$$

It is a scalar quantity, that is it is invariant under Galilean transformations (including rotations). The scalar product of two different three vectors is also an invariant:

$$\mathbf{r}_1 \cdot \mathbf{r}_2 = x_1 x_2 + y_1 y_2 + z_1 z_2.$$

4   The **inner product** of a four-vector x with itself is defined by

$$x \cdot x = (x^0)^2 - (x^1)^2 - (x^2)^2 - (x^3)^2,$$
$$= (ct)^2 - \mathbf{r} \cdot \mathbf{r}, \tag{8.36}$$

where the superscript outside the brackets is an exponent. We know from equation (7.3) that the right-hand side of the last line is an invariant under Lorentz transformations.

5   The inner product of two different four-vectors x and y is

$$x \cdot y = x^0 y^0 - x^1 y^1 - x^2 y^2 - x^3 y^3 \tag{8.37}$$

and is also an invariant. The reader should confirm this fact by completing Problem 8.3. Note that $x \cdot y = y \cdot x$.

6   Look out for the confusion of terms: A *scalar quantity* means an *invariant quantity* (means a *rank zero tensor*!).

7   We shall frequently use the notation $x \cdot x = x^2$ when there is no possibility of confusion with x^2 meaning the component of x labelled with a superscript 2.

8   We shall pick up this development of four-vectors later (Section 9.2). Now we must become accustomed to using them. And we start by finding some others.

In Section 7.16 we introduced the concept of *proper time*. In simple terms it is the time, τ, on a clock that moves with a particle. More formally, the infinitesimal interval on this clock $d\tau$ between two events on the world line of this particle is an invariant. From any other inertial frame Σ it may be calculated using equation (7.15):

$$d\tau = dt\sqrt{1 - v^2/c^2} = dt/\gamma, \tag{8.38}$$

where dt is the observed time interval between those two events and v is the instantaneous speed of that particle, both as observed from Σ. We now use the proper time to form some more four-vectors.

a. *The four-vector of velocity.* Suppose now that the four-vector x represents the coordinates of a particle moving in Σ. The vector x can be considered as a function of the proper time τ. Since $d\tau$ is an invariant the object $dx/d\tau$ is also a four-vector. Explicitly:

$$\frac{dx}{d\tau} = \left(\frac{dx^0}{d\tau}, \frac{dx^1}{d\tau}, \frac{dx^2}{d\tau}, \frac{dx^3}{d\tau} \right)$$
$$= \left(\frac{dct}{d\tau}, \frac{d\mathbf{r}}{d\tau} \right)$$
$$= \left(\frac{c\,dt}{d\tau}, \frac{d\mathbf{r}}{dt}\frac{dt}{d\tau} \right).$$

But from equation (8.38) we have

$$\frac{dt}{d\tau} = \gamma = \frac{1}{\sqrt{1 - v^2/c^2}}$$

and the particle's velocity $\mathbf{v}$ is

$$\mathbf{v} = \frac{d\mathbf{r}}{dt}.$$

Therefore

$$\frac{dx}{d\tau} = (\gamma c, \gamma \mathbf{v}) = c(\gamma, \gamma \boldsymbol{\beta}) \tag{8.39}$$

is a four-vector. This is the velocity four-vector, or **four-velocity**. We anticipated the existence of this four-velocity in Section 7.15.

The fact that we involved a particle in this discussion is relevant. The velocity of a particle or of the centre-of-mass of a group of particles, or the *group velocity* (Section 9.3) of a wave, may all be expressed as a four-vector as in equation (8.39). However, in general the velocities associated with waves must be treated differently. We start to consider this matter when we introduce the wave four-vector in (e) of this section.

Recalling equations (7.17)–(7.19) in order to transform a velocity is not easy. Remembering the correct four-vector makes transforming a velocity a simple matter. We shall see examples of this in Section 8.12.

We have used the symbol P for momentum and shall use p for four-momentum. Since we downgraded speed to v and velocity to $\mathbf{v}$, we cannot use v for four-velocity. Instead we shall use $\bar{u}, \bar{v}, \bar{w}, \ldots$ for four-velocities, where $\bar{u} = (\gamma c, \gamma \mathbf{u})$, $\gamma = 1/\sqrt{1 - u^2/c^2}$, and so on. Sometimes we shall use the fractional four-velocity $\bar{u}/c = (\gamma, \gamma \boldsymbol{\beta})$, $\boldsymbol{\beta} = \mathbf{u}/c$, $\gamma = 1/\sqrt{1 - \beta^2}$, and so on. To prevent a velocity becoming lost in the anonymity of γ and β we may add subscripts to these symbols: β_u for u/c, and so on.

Problem 8.4 asks the reader to show that for any four-velocity $\bar{u}$, $\bar{u} \cdot \bar{u} = c^2$.

b. *The energy–momentum four-vector.* Take the fractional four-velocity $\bar{v}/c = (\gamma, \gamma \boldsymbol{\beta})$ of a particle. Form another four-vector p by multiplying each component by the rest mass energy of the particle. Thus

$$p = (\gamma Mc^2, \gamma \boldsymbol{\beta} Mc^2)$$

$$= \left(\frac{Mc^2}{\sqrt{1 - v^2/c^2}}, \frac{M\mathbf{v}c}{\sqrt{1 - v^2/c^2}} \right).$$

From Table 2.2 we can write this as

$$p = (E, \mathbf{P}c), \tag{8.40}$$

where E is the particle's total energy and $\mathbf{P}$ is its momentum. We are aware already that $(E, \mathbf{P}c)$ is a four-vector because the transformation we quoted in Section 6.2

for these quantities is the same as that for the time–space coordinates $x = (ct, \mathbf{r})$, as stated explicitly in Section 7.4. But now we understand why.

We can call p the (energy-)momentum four-vector (or **four-momentum**) and we see that $p \cdot p$ is another way of writing the invariant $E^2 - \mathbf{P} \cdot \mathbf{P}c^2 = (Mc^2)^2$. And, of course, the inner product of two energy and momentum four-vectors $p_1 \cdot p_2$ is also an invariant.

Readers will recall that it was Planck who first defined (Section 5.9) the momentum of a particle as given by

$$\mathbf{P} = \frac{M\mathbf{v}}{\sqrt{1 - v^2/c^2}}.$$

In Chapter 3 we assumed that this $\mathbf{P}$ and E were the particle quantities which contributed to the overall conserved momentum and energy in collisions between particles. That this was the case shown by Lewis and Tolman in 1909 (see Section 5.9) and we give an outline of their demonstration in Section 8.12. That means we can now think of collisions as conserving the total four-vector of momentum. (See Section 9.2.)

A photon has no proper time and no four-velocity so that we cannot obtain a four-momentum for a photon by the argument just applied to particles with rest mass. The result that the four-momentum is $(E, \mathbf{P}c)$ where $E = |\mathbf{P}|c$ is implicit in Table 9.1 and in the discussion of Section 9.4.

c. *The acceleration four-vector.* The acceleration of a particle is the time rate of change of velocity. If we want to construct a four-vector we must use the four-velocity, $\bar{v}$ say, and the proper time τ. Thus the four-acceleration $\bar{a}$ is given by

$$\bar{a} = \frac{d\bar{v}}{d\tau}. \tag{8.41}$$

In the rest frame of the particle, a has the components $(0, \mathbf{a})$ where $\mathbf{a}$ is the Newtonian acceleration. The case of constant $\mathbf{a}$ in the rest frame allows a simple integration of the equations of motion of the particle in the frame in which the particle was at rest at some convenient time. The result gives distance travelled in terms of time in that frame and the elapsed proper time. See Problems 8.4–8.8.

d. *The Minkowski force.* An acceleration suffered by a particle means a force acting on that particle. We can define the force acting on a particle of rest mass M that experiences a four-acceleration $\bar{a}$ by

$$\bar{f} = M\bar{a}. \tag{8.42}$$

Since M is an invariant $\bar{f}$ is clearly a four-vector. This is a Minkowski force.

e. *The wave four-vector.* A travelling, sinusoidal plane wave may be characterised by two numbers: the wavelength λ and the frequency f. The phase velocity is then λf. The displacement associated with a simple wave might have a time and space dependence represented by a function of time t and distance x (in the direction of wave travel, not the four-vector) by a function such as $\sin 2\pi(ft - x/\lambda)$.

More convenient are the wave number $\chi = 2\pi/\lambda$ and angular frequency $\omega = 2\pi f$ in which case the phase velocity is ω/χ and the displacement has a time and space dependence given by $\sin(\omega t - \chi x)$. We call $\omega t - \chi x$ the phase of the wave. If this plane wave is moving in three space dimensions then χ is replaced by a three-vector $\mathbf{k}$, pointing in a direction normal to the wave front (Figure 8.7). For any point on the wave front with coordinates $\mathbf{r}$, the scalar $\mathbf{k} \cdot \mathbf{r}$ always has the same value. Then, at a fixed time, the phase is also the same across the wavefront, as required. In relativity the phase should be an invariant since, for example, the wave reaching a maximum displacement is an event not tied to a particular inertial reference frame. We define the **wave four-vector** by

$$k = (\omega/c, \mathbf{k}). \tag{8.43}$$

Then, remembering that $x = (ct, \mathbf{r})$, we see that

$$k \cdot x = \omega t - \mathbf{k} \cdot \mathbf{r}$$

is the invariant phase required. And we have discovered another four-vector.

We note that physicists are happier employing the function $\exp(ik \cdot x)$ to represent the space–time variation of a plane wave. And that, since the four-vector x has dimensions of length for all components, all the components of k must and do have the dimension of $(\text{length})^{-1}$.

We continue discussing waves in the context of wave–particle duality in Sections 9.3 and 9.4.

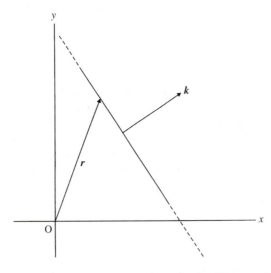

Figure 8.7 A two-dimensional sketch of a plane wavefront at a fixed time. The wave three-vector $\mathbf{k}$ is in the direction of the normal to the wavefront and has a magnitude $2\pi/\lambda$, where λ is the wavelength. The vector $\mathbf{r}$ is the coordinate of a point on the wavefront. It is clear that $\mathbf{k} \cdot \mathbf{r}$ has the same value at all points on the wavefront.

*8.11 A classification of four-vectors

In Section 7.6 we introduced the idea of the squared interval between events (equation (7.6)):

$$(\Delta s)^2 = (\Delta ct)^2 - (\Delta x)^2 - (\Delta y)^2 - (\Delta z)^2.$$

We defined three types of interval and discussed the physics of the relation between the two events involved. The intervals had:

1 $(\Delta s)^2 = 0$ One event is on the light cone of the other.
2 $(\Delta s)^2 > 0$ Time-like interval.
3 $(\Delta s)^2 < 0$ Space-like interval.

But the interval $\Delta s = (\Delta ct, \Delta \mathbf{r})$ is a four-vector, so that $(\Delta s)^2 = \Delta s \cdot \Delta s$. And, therefore, any four-vector may be classified in a similar way:

1 A null four-vector V has $V \cdot V = 0$.
2 A time-like four-vector V has $V \cdot V > 0$.
3 A space-like four-vector V has $V \cdot V < 0$.

Of course, $V \cdot V$ is an invariant and, as in the case of Δs, no member of one class may be transformed into another class. In each case a transformation can be found that reduces V to a simple form. These forms are not unique but have the structure given by the following statements:

1 A null vector V can be transformed to the form $(V^0, 0, 0, V^0)$.
2 A time-like vector V can be transformed to the form $(V^0, 0, 0, 0)$.
3 A space-like vector V can be transformed to the form $(0, 0, 0, V^0)$.

The important point here is that the time-like vectors can be transformed to have no space components, and the space-like vectors can be transformed to have no time component. The non-uniqueness concerns the space component in 1 and 3. It does not have to be in the z direction, but can be along any axis, or in fact along any space direction.

+8.12 Relativistic energy and momentum conservation validated

We have assumed that energy and momentum are conserved in particle collisions and decays. The relativistic energy E and momentum $\mathbf{P}$ that we assigned to each particle involved is that given in Table 2.2. Namely

$$E = \frac{Mc^2}{\sqrt{1 - v^2/c^2}}, \quad \mathbf{P}c = \frac{M\mathbf{v}c}{\sqrt{1 - v^2/c^2}}, \tag{8.44}$$

where $\mathbf{v}$ is the velocity of the particle in the inertial frame in which the process is considered, and M is its rest mass. We wish now to confirm that these quantities are appropriate to energy and momentum conservation in collisions. This was first done by G. N. Lewis and R. C. Tolman in 1909,[3] three years after Planck defined the relativistic momentum (equation (5.19), Section 5.9). Einstein's first postulate (Section 5.5) means that if energy and momentum are conserved in one inertial frame they must be conserved in any other inertial frame. Lewis and Tolman imagined a collision in which momentum was obviously conserved. They transformed the velocities to another frame and asked how momentum should depend on speed if momentum was conserved in this new frame. We shall follow their method.

Their argument naturally assumed the correctness of the Lorentz transformations of velocity found by Einstein. We assume we can use the four-velocity to transform three-vectors of velocity. Thus a three-velocity $\mathbf{v}$ has a *fractional four-velocity* $\bar{v}/c$ given by

$$\bar{v}/c = (\gamma_v, \gamma_v \boldsymbol{\beta}_v), \tag{8.45}$$

where, as usual, $\boldsymbol{\beta}_v = \mathbf{v}/c$ and $\gamma_v = 1/\sqrt{1 - \beta_v^2}$. Conversely, if we find what we think is a four-vector of velocity $\bar{v}/c$ with components $(G, \mathbf{B})$ then we can confirm by checking that it satisfies

$$\bar{v} \cdot \bar{v}/c^2 = G^2 - \mathbf{B} \cdot \mathbf{B} = 1. \tag{8.46}$$

Then we can interpret this to have a three-velocity $\mathbf{v}$ given by

$$\mathbf{v}/c = \mathbf{B}/G. \tag{8.47}$$

Note that we are frequently going to use the fractional four-velocity $(\gamma, \gamma\beta)$ in place of the four-velocity $(\gamma c, \gamma \boldsymbol{\beta} c) = c(\gamma, \gamma \boldsymbol{\beta})$.

Step 1. We consider the elastic collision between two equal mass particles, J and K (Figure 8.8). We use the standard arrangement of two inertial reference frames Σ and Σ'. In Σ, J moves, with speed v, down the x-axis towards the origin O of Σ where it collides with K at $t = 0$. The collision is elastic and such that J recoils with speed v along the x-axis. In Σ' K moves towards O' on the $-x'$-axis with speed v and meets J at O' at time $t' = 0$. The collision is such that K recoils with speed v along the $-x'$-axis. The observers in Σ and in Σ' agree that J in Σ suffers a collision identical to that suffered by K in Σ', and that in each frame the momentum change is the same in magnitude as in the other frame.

Step 2. Now we investigate how an observer in Σ sees the motion of both particles. To do this we must transform the velocity of K in Σ' to its value in Σ. The four-velocity of K before the collision has components $c(\gamma_v, -\gamma_v\beta_v, 0, 0)$ and after the collision has components $c(\gamma_v, +\gamma_v\beta_v, 0, 0)$. As usual we take Σ' to be moving with speed u in the direction of the z-axis in Σ. The matrix transforming

---

3 Lewis, G. N. and Tolman, R. C., *Philosophical Magazine* **18**, 1909, 510–23.

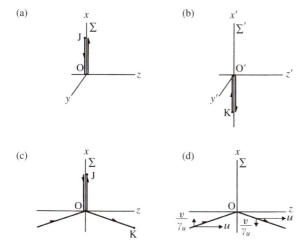

Figure 8.8 The collision between two equal mass particles J and K in our standard arrangement of inertial reference frames. (a) In Σ, J is moving down the x-axis towards the origin O with speed v. At the origin, at time $t = 0$, it collides with K in such a way that its velocity is exactly reversed. (b) In Σ′, K is moving on the $-x'$-axis towards the origin O′ with speed v. At this origin, at time $t' = 0$, it collides with J in such a way that its velocity is exactly reversed. (c) The collision as it appears to an observer in Σ allowing for the motion of Σ′ in Σ. (d) Shows the components of the three-velocity of K in Σ before and after the collision.

these four-vectors from Σ′ to Σ is equation (7.5)

$$
\mathbf{L}(\beta_u) =
\begin{pmatrix}
\gamma_u & 0 & 0 & \gamma_u \beta_u \\
0 & 1 & 0 & 0 \\
0 & 0 & 1 & 0 \\
\gamma_u \beta_u & 0 & 0 & \gamma_u
\end{pmatrix}.
$$

We leave it to the reader to check the results of the matrix multiplication. The following Table 8.2 shows the results for K along with the values of the components of the vectors of the velocity of J in Σ, before and after the collision for both particles. Note that the table entries are fractional velocities.

Step 3. We examine the conservation of the x-component of momentum in Σ. If we use the Newtonian result *momentum = mass × velocity* we see that J increases the x-component of its momentum by $2Mc\beta_v$ where as K decreases the x-component of its momentum by $2Mc\beta_v/\gamma_u$. Thus momentum is not conserved or we have the wrong momentum. To proceed we suppose the momentum is given by an alternative expression:

$$
\mathbf{P} = Mf(v/c)\mathbf{v}, \tag{8.48}
$$

Table 8.2 Fractional velocities in Σ

Particle	J		K	
Component	Before	After	Before	After
Four-velocities/c in Σ				
0	γ_v	γ_v	$\gamma_u\gamma_v$	$\gamma_u\gamma_v$
1	$-\gamma_v\beta_v$	$+\gamma_v\beta_v$	$+\gamma_v\beta_v$	$-\gamma_v\beta_v$
2	0	0	0	0
3	0	0	$\gamma_v\gamma_u\beta_u$	$\gamma_v\gamma_u\beta_u$
Three-velocities/c in Σ				
x	$-\beta_v$	$+\beta_v$	$+\beta_v/\gamma_u$	$-\beta_v/\gamma_u$
y	0	0	0	0
z	0	0	β_u	β_u
Magnitude of three-velocity/c in Σ				
	β_v	β_v	$\sqrt{\beta_u^2 + (\beta_v/\gamma_u)^2}$	$\sqrt{\beta_u^2 + (\beta_v/\gamma_u)^2}$

where $f(v/c)$ is a function of the speed v with the property that the limit of $f(v/c)$ as $v/c \to 0$ is 1 in order to match the Newtonian result at low speeds. This function must also be a scalar in order that the vector of this momentum is in the same direction as the velocity. Then the conservation of the x-component of momentum requires

$$2Mf(\beta_v)\beta_v c = 2Mf\left(\sqrt{\beta_u^2 + (\beta_v/\gamma_u)^2}\right)\beta_v c/\gamma_u$$

or

$$\gamma_u f(\beta_v) = f\left(\sqrt{\beta_u^2 + (\beta_v/\gamma_u)^2}\right), \tag{8.49}$$

where $\beta_v = v/c$, $\beta_u = u/c$, and $\gamma_u = 1/\sqrt{1 - \beta_u^2}$. The function that satisfies equation (8.49) is

$$f(\beta_v) = f(v/c) = 1/\sqrt{1 - v^2/c^2}.$$

Thus the conserved momentum (equation (8.48)) is

$$\mathbf{P} = \frac{M\mathbf{v}}{\sqrt{1 - v^2/c^2}}. \tag{8.50}$$

Step 4. What about energy? In the collision that we have described there is no transfer of energy and we have no handle on this quantity. However, we can transform into an inertial frame Σ'' in which J is initially at rest. Our specification of Σ'' is that to an observer in Σ'' the course of the same collision is that at time $t'' = 0$ the stationary J is struck by K and recoils along the x''-axis. See Figure 8.9. The transformation from Σ to Σ'' is similar to our standard arrangement: Σ'' is moving with speed v along the $-x$-axis in Σ. The origins and axes of the two

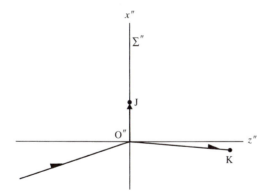

Figure 8.9 The J–K collision as observed in the frame Σ″. J is initially at rest at the origin O″ and is struck at time $t'' = 0$ by K so that it recoils along the x''-axis.

Table 8.3 Fractional velocities in Σ″

Particle	J		K	
Component	Before	After	Before	After
Four-velocities/c in Σ″				
0	1	$\gamma_v^2 + \gamma_v^2\beta_v^2$	$\gamma_u\gamma_v^2 + \gamma_v^2\beta_v^2$	$\gamma_u\gamma_v^2 - \gamma_v^2\beta_v^2$
1	0	$2\gamma_v^2\beta_v$	$\gamma_u\gamma_v^2\beta_v + \gamma_v^2\beta_v$	$\gamma_u\gamma_v^2\beta_v - \gamma_v^2\beta_v$
2	0	0	0	0
3	0	0	$\gamma_v\gamma_u\beta_u$	$\gamma_v\gamma_u\beta_u$

frames coincide at $t'' = 0 = t$. The matrix that, acting on a four-vector in Σ, produces that four-vector's components in Σ″ is:

$$\begin{pmatrix} \gamma_v & \gamma_v\beta_v & 0 & 0 \\ \gamma_v\beta_v & \gamma_v & 0 & 0 \\ 0 & 0 & 1 & 0 \\ 0 & 0 & 0 & 1 \end{pmatrix}.$$

Operate on all the J and K four-vectors in Table 8.2. Table 8.3 shows the results for four-velocities before and after the collision. As required, the result of this transformation has left J at rest in Σ″ before the collision.

Step 5. In Table 2.2 we assigned to a particle rest mass M, speed v, an energy E given by

$$E = \frac{Mc^2}{\sqrt{1 - v^2/c^2}} = Mc^2\gamma_v.$$

Now we remind the reader that a particle velocity $\mathbf{v} = \boldsymbol{\beta}_v c$ has a fractional four-velocity that is $(\gamma_v, \gamma_v \boldsymbol{\beta}_v)$. Therefore, as in equation (8.40), this energy assignment is

Energy = rest mass energy

× zeroth component of its fractional four-velocity.

If we look at the tabulated quantities above (Table 8.3), we can pick out the zeroth components of the four-velocities and write down the sum of the particle energies before and the sum after the collision:

The $J + K$ energy sum before $= Mc^2 + Mc^2(\gamma_u \gamma_v^2 + \gamma_v^2 \beta_v^2)$

The $J + K$ energy sum after $= Mc^2(\gamma_v^2 + \gamma_v^2 \beta_v^2) + Mc^2(\gamma_u \gamma_v^2 - \gamma_v^2 \beta_v^2)$.

These two are the same (reader check!). Therefore our energy assignment is consistent with energy conservation in this collision.

Summary. There are some points to be made about these arguments. In what follows, the discussion focuses on the energy and momentum of a single particle. The conservation discussed concerns the totality of these quantities in particle collisions and decays. In that case every particle involved has its own energy and momentum to be counted in an appropriate way.

1   The discussion was based on the behaviour of the four-vectors of velocity under Lorentz transformations. Such a four-vector $(\gamma c, \gamma \boldsymbol{\beta} c)$ is multiplied by Mc in order to obtain the energy and momentum $(E, \mathbf{P}c)$ of a particle with this four-velocity. But we would still have a conservation of the quantity that was the four-velocity multiplied by αMc where $\alpha \neq 1$. This quantity would not identify with what we had previously taken to be the energy and momentum. However, if we impose the requirement that the relativistic momentum $\mathbf{P}$ has the same value as the Newtonian momentum $M\mathbf{v}$ in the limit of $v/c \to 0$, then we cannot avoid $\alpha = 1$.

2   The relativistic energy E does not go to the Newtonian kinetic energy $\frac{1}{2}Mv^2$ in the limit $v/c \to 0$, but to $Mc^2 + \frac{1}{2}Mv^2$ (Section 2.2). The rest mass energy Mc^2 has to be included in E if we want to have a four-vector of energy and momentum. Since the particles J and K had equal mass and did not change rest mass, the conservation of energy that we have discussed could be thought only to be conservation of kinetic energy. The possibility that rest mass energy could become kinetic energy or kinetic energy could become rest mass was not considered.

3   The evidence from nuclear and particle physics tells us that rest mass can be transformed into kinetic energy, and the other way (Sections 2.5–2.7). That suggests that it is the rest mass energy plus kinetic energy that is conserved. And that is E, the time (zeroth) component of the energy–momentum four-vector.

4   There is a simple argument about the conservation of energy. Suppose that a group of particles is not subjected to any external forces and that there is a collision or a decay. If $(E_i, \mathbf{P}_i c)$ is the initial and $(E_f, \mathbf{P}_f c)$ is the final total four-momentum and we assume the conservation of this energy and momentum then $(E_i - E_f, \mathbf{P}_i c - \mathbf{P}_f c)$ is a four-vector with all components zero (Section 8.10). Such a four-vector has the same property in all inertial frames. Therefore, given that special relativity requires that a physics law is true in all inertial frames, it follows that the law of conservation of momentum $\mathbf{P}_f - \mathbf{P}_i = 0$ must mean that the energy that is conserved in all frames is E, the time (zeroth) component of the four-momentum as we have used it.

Thus we conclude that the assignment of energy and momentum as in equation (8.44) is consistent with energy and momentum conservation at all attainable particle speeds and, of course, in all inertial frames.

+8.13   The Thomas precession

Precession in physics is a rotation of the axis of a spinning body about another axis due to a torque acting to change the direction of the first axis. Many readers will be familiar with the action of a toy gyroscope. The axis of rapid rotation of the gyroscope's heavy disc can be made to precess around the vertical. The torque is provided by gravity acting down at the disc centre and a reaction at the point of support of the disc on its axis.

We start with a Bohr-like model (Figure 8.10(a)) of an electron moving in a circular orbit around an atomic nucleus. We assume that this electron has intrinsic angular momentum (spin) which acts as does a gyroscope. In the absence of any couple acting on the spin, we expect that as the electron moves around its orbit

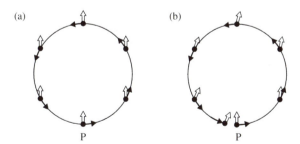

Figure 8.10 (a) This diagram shows the non-relativistic behaviour of the spin vector (⇑) of an electron as it traverses a single turn of a circular, classical orbit around an atomic nucleus. (b) Shows how, if the electron speed around the orbit is not vanishingly small compared to c, the spin vector turns in the opposite sense to the rotation of the electron. This is the Thomas precession.

the direction of this gyroscope's axis of rotation (= direction of the spin) remains unchanged with respect to the fixed stars. So after making one complete revolution from the point P the electron is again at P with its spin oriented in the same direction with respect to its instantaneous velocity as when it was last at P. This is correct if the magnitude of the velocity v is negligibly small compared with c.

However, at velocities not vanishingly small compared to c this is not the case. The spin direction rotates in the opposite sense to the rotation around the orbit. After traversing one complete orbit the spin will no longer point in the same direction as it did before this single orbit (see Figure 8.10(b)). Thus as the electron traverses many orbits, the electron's spin vector rotates in the reference frame of the nucleus in a sense opposite to the rotation of the electron's velocity vector. Thus this spin rotation is around a direction perpendicular to the orbit. This is the Thomas precession.

We must stress that the Thomas precession is a geometrical effect. There is no mysterious couple acting on the electron that causes the spin angular momentum vector to precess in this direction. The cause is related to the fact that Lorentz transformations do not in general commute (Section 8.9) and that when a particle is accelerated in a direction not along the direction of its velocity there is a rotation of reference rest frames. The electron spin direction is the property affected by this rotation. However, there is a couple acting on the electron's spin. The electron is moving in the electric field of the nucleus. In the rest frame of the electron that electric field has a magnetic part (Section 9.8) which, acting on the magnetic moment of the electron, causes its spin to precess in the opposite sense to the Thomas precession. The total effect of 'magnetic' less 'Thomas' is equal to half 'magnetic' alone. Without 'Thomas', the then existing atomic theory (1926) predicted the fine structure splitting of some atomic spectral lines to be twice that observed. With Thomas precession included, theory and experiment came into agreement.

The difficulty is that the electron's rest frame is not an inertial frame. The electron is accelerating towards the centre of its orbit. We have to make use of what we call Instantaneously Co-moving Inertial Frames (ICIFs). Consider an electron in a circular orbit about an atomic nucleus which we assume to be stationary in the laboratory. At any one instant it is possible to find a Lorentz boost that transforms from the laboratory to the rest frame of the electron. That is an ICIF to the electron. We assume that at that instant the properties of the electron are described as if it was in that ICIF. Now let us find the ICIFs we need.

Suppose the electron is moving around a circular orbit of radius r with speed βc. In a short time Δt the electron moves from P_1 to P_2 and the velocity vector βc turns through an angle $\beta c \Delta t / r$. The electron's velocity changes from βc to $\beta c + \Delta \beta c$ (Figure 8.11). The laboratory reference frame is called Σ and has its z-axis tangent to the orbit at P_1. For convenience and with no loss of generality, the orbit lies in

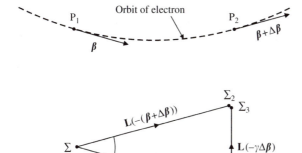

Figure 8.11 This figure shows the velocity vectors of an electron at point P_1 and at a nearby point P_2 on its orbit. The triangle symbolises the Lorentz transformations that connect the laboratory frame Σ and the frame Σ_1 in the ICIF1 and the frames Σ_2 and Σ_3 in the ICIF2. (This diagram has exaggerated a distance between P_1 and P_2. In a proper analysis, the rate of precession is found in the limit Δt (and $\Delta\beta$) $\rightarrow$ 0.)

the yz plane. Evidently the vectors $\boldsymbol{\beta}c$ and $\boldsymbol{\beta}c + \Delta\boldsymbol{\beta}c$ must also lie in this plane. The two ICIFs that we need are:

ICIF1. This is ICIF to the electron at P_1 at laboratory time t. A reference frame Σ_1 in this inertial frame is reached by a boost $\mathbf{L}(-\boldsymbol{\beta})$ from Σ. (We represent a boost by the matrix $\mathbf{L}(\boldsymbol{\beta})$ as in Section 6.2 and subsequently, particularly 8.9. Remember that $\mathbf{L}$ operates on a column matrix to its right.)

ICIF2. This is an ICIF to the electron at P_2 at laboratory time $t + \Delta t$. However, there are two relevant reference frames in this ICIF:

i Σ_2 reached by a boost $\mathbf{L}(-(\boldsymbol{\beta} + \Delta\boldsymbol{\beta}))$ from Σ,
ii Σ_3 reached by a boost $\mathbf{L}(-\gamma\Delta\boldsymbol{\beta})$ from Σ_1.

(The γ factor is present in boost (ii) because the velocity $\Delta\boldsymbol{\beta}$ is defined in the laboratory but has to be $\gamma\Delta\boldsymbol{\beta}$ in Σ_1.) Now there are two ways of transforming from ICIF1 to ICIF2:

a $\Sigma_1 \rightarrow \Sigma \rightarrow \Sigma_2$ generated by $\mathbf{L}(-(\boldsymbol{\beta} + \Delta\boldsymbol{\beta})) \, \mathbf{L}(+\boldsymbol{\beta})$,
b $\Sigma_1 \rightarrow \Sigma_3$ generated by $\mathbf{L}(-\gamma\Delta\boldsymbol{\beta})$.

But we know from Section 8.9 that two successive Lorentz transformations (a) are equivalent to a single transformation (b) *and* a rotation. Therefore $\Sigma_2 \neq \Sigma_3$ and Σ_3 must be rotated around the axis perpendicular to the electron's orbit to bring it into coincidence with Σ_2.

Now the direction of the electron's spin with respect to the axes in Σ_1 is carried unchanged into Σ_3 under (b). Σ_3 is rotated with respect to Σ_2. The result is that, observed from the laboratory, the electron's spin has rotated in the opposite sense to the rotation of the electron's velocity vector. Over a finite time the spin precesses as in Figure 8.10(b), in the absence of any other spin changing effects. The angular velocity $\boldsymbol{\omega}$ of the spin in the frame in which the nucleus is at rest is given by

$$\boldsymbol{\omega} = \frac{\gamma}{(\gamma + 1)c^2}\mathbf{a} \times \mathbf{v},$$

where $\mathbf{v}$ is the electron's velocity and $\mathbf{a}$ its acceleration. Thus in Figure 8.10 $\mathbf{a}$ is directed towards the centre of the circle so that $\boldsymbol{\omega}$ is a vector pointing into the plane of the paper and representing a clockwise rotation of the electron spin.

This has been a somewhat simplified account of the Thomas precession. Derivations of the final formula for the Thomas precession are given by, amongst others, Goldstein,[4] and Jackson.[5] Of course, a Bohr model of the atom containing planet-like electron orbits around a nucleus is incorrect. The next step was taken by Schrödinger. Solutions of his equation (Section 9.5) for a single electron moving in the Coulomb field of a proton gave the principal features of the hydrogen atom energy levels. But the fine structure of the energy levels was not given by Schrödinger's wave mechanics. They could only be explained by arbitrarily including the effect of both the electron spin and the Thomas precession.[6] This is interesting as Schrödinger's theory has no planet-like electron orbit although that is the context in which the Thomas precession is explained. Dirac's relativistic equation (Section 9.6) solved for the electron in a hydrogen atom gives the principal energy levels and their fine structure correctly without the arbitrary addition of electron spin and the Thomas precession.

The Thomas precession of the spin applies to all particles with spin experiencing an acceleration with a component perpendicular to their velocity. The complete relativistic treatment of spin motion for particles moving in constant or slowly changing electric and magnetic fields may be found in Jackson.[7]

Problems

8.1   The answer to Problem 7.9 may be found by a couple of Lorentz transformations of the time and space coordinates of events. The first is from A's rest frame to that of B of the event at which B receives the former's light signal. The second transforms to A's rest frame the coordinates of the event in B's rest frame $2T$ seconds later when B sends an acknowledgement. Then add the time for the light to travel back to A. Did you find an easier method to

4 Goldstein, H., *Classical mechanics*, 2nd edn, Reading, Mass.: Addison Wesley, 1980.
5 Jackson, J. D., *Classical electrodynamics*, 2nd edn, New York: John Wiley & Sons, 1975.
6 Woodgate, G. K., *Elementary atomic structure*, 2nd edn, Oxford: Oxford University Press, 1980, p. 62.
7 Jackson, J. D., *op. cit.*, pp. 556–60.

find the answer? If not, try using the k-calculus (Section 8.2, step 2), or the formula for the Doppler effect.

8.2 The equations for a general Lorentz boost with velocity $\mathbf{u} = \boldsymbol{\beta}_u c$ of the time–space coordinates of an event may be written

$$ct = \gamma_u(ct' + \boldsymbol{\beta}_u \cdot \mathbf{r}'),$$

$$\mathbf{r} = \mathbf{r}' + (\gamma_u - 1)(\boldsymbol{\beta}_u \cdot \mathbf{r}')\boldsymbol{\beta}_u / \beta_u + \gamma_u \boldsymbol{\beta}_u ct',$$

where, as usual $\gamma_u = 1/\sqrt{1 - \beta_u^2}$. Show that these equations reduce to the transformation equations for the coordinates of an event in the standard arrangement (equations (7.4)) if $\mathbf{u}$ has $u_x = u_y = 0$ and $u_z = u$.

8.3 Show that the inner product of any two four-vectors is an invariant.

8.4 If $\bar{v}$ is a four-velocity of a particle show that $\bar{v} \cdot \bar{v} = c^2$ and that in the rest frame of the particle $\bar{v}$ has one non-zero component, $\bar{v}^0$. Given that the particle has a four-acceleration $\bar{a} = d\bar{v}/d\tau$ (Section 8.10) where τ is the proper time, show that $\bar{a} \cdot \bar{v} = 0$.

8.5 Consider the problem of the smooth acceleration of a particle along one direction, x say. Since this is a two-dimensional problem every four-vector has only two non-zero components. Thus instead of four-velocity $\bar{v} = (\bar{v}^0, \bar{v}^1, \bar{v}^2, \bar{v}^3)$ we write $\bar{v} = (\bar{v}^0, \bar{v}^1) = (\gamma_v c, \gamma_v v)$, and similarly for the four-acceleration $\bar{a} = (\bar{a}^0, \bar{a}^1)$. We assume that in the rest frame of the particle the acceleration is constant and, as an example, equal to that of gravity, g, at the surface of the Earth. To obtain the components of $\bar{a}$ in the rest frame, differentiate $\bar{v} = (\bar{v}^0, \bar{v}^1) = (\gamma_v c, \gamma_v v)$, to obtain the general $\bar{a}$ and then put $v = 0$ to obtain its rest frame components $(\bar{a}^0, \bar{a}^1)_{RF}$:

$$(\bar{a}^0, \bar{a}^1)_{RF} = \left(\frac{d\bar{v}^0}{d\tau}, \frac{d\bar{v}^1}{d\tau}\right)_{v=0} = \left(0, \frac{dv}{d\tau}\right) = (0, g).$$

It follows that the invariant $\bar{a} \cdot \bar{a} = (\bar{a}^0 \bar{a}^0 - \bar{a}^1 \bar{a}^1)_{RF} = -g^2$. We now have the three equations that are valid in all inertial frames:

$$\bar{v} \cdot \bar{v} = \bar{v}^0 \bar{v}^0 - \bar{v}^1 \bar{v}^1 = c^2,$$

$$\bar{a} \cdot \bar{v} = \bar{a}^0 \bar{v}^0 - \bar{a}^1 \bar{v}^1 = 0, \text{ and}$$

$$\bar{a} \cdot \bar{a} = \bar{a}^0 \bar{a}^0 - \bar{a}^1 \bar{a}^1 = -g^2.$$

By eliminating $\bar{a}^0$ from the second and third equations, show that

$$(\bar{a}^1)^2 = g^2 (\bar{v}^0)^2 / c^2.$$

Since the problem posed has the accelerations in the direction of increasing x, we can take the positive roots of this equation and find

$$\bar{a}^1 = \frac{d\bar{v}^1}{d\tau} = g\bar{v}^0 / c.$$

Show that

$$\bar{a}^0 = \frac{d\bar{v}^0}{d\tau} = \frac{g\bar{v}^1}{c}.$$

Integrate these coupled equations to find expressions for $\bar{v}^0$ and $\bar{v}^1$ as functions of the proper time τ in the case of the particle starting from rest in a given frame Σ (initial conditions are $\tau = 0$, $\bar{v}^0 = c$, and $\bar{v}^1 = 0$). Show that at the proper time τ for the particle, the non-zero components its four-velocity in Σ are

$$\bar{v}^0 = c \cosh\frac{g\tau}{c} \quad \text{and} \quad \bar{v}^1 = c \sinh\frac{g\tau}{c}.$$

8.6   Starting with the final result of the last question and the fact that in Σ we have $\bar{v}^1 = dx/d\tau$ and $\bar{v}^0 = c(dt/d\tau)$, integrate and use the initial conditions $t = 0$ and $x = 0$ at $\tau = 0$ to find formulae for t as a function of τ, and for x and v ($= dx/dt$) as a function of t and of τ. Thus show that

$$1 \quad t = \frac{c}{g}\sinh\frac{g\tau}{c},$$

$$2 \quad x = \frac{c^2}{g}\left(\cosh\frac{g\tau}{c} - 1\right) = \frac{c^2}{g}\left\{\sqrt{1 + \left(\frac{gt}{c}\right)^2} - 1\right\}, \quad \text{and}$$

$$3 \quad v = \frac{dx}{dt} = c\tanh\frac{g\tau}{c} = \frac{gt}{\sqrt{1 + (gt/c)^2}}$$

8.7   Consider an inertial frame Σ in which two rockets are at rest a distance d apart on the x-axis. At $t = 0$ they set their proper clocks to zero and commence identical acceleration programmes in the direction of increasing x. As in Problem 8.5 the acceleration is g in the rest frame of each rocket. Show that at a time t in Σ

1   the proper times are the same,
2   the speeds measured in Σ are the same,
3   measured in Σ the rockets are the same distance d apart as when they started moving.

Show also that in the inertial frame instantaneously co-moving with the rockets they are a distance apart $d/\sqrt{1 - v^2/c^2}$ where v is the instantaneous speed in Σ.

8.8   The starship *Enterprise* is ordered to the centre of the Galaxy which is believed to be a distance of 25,000 light years from the Earth. The *Enterprise* can sustain an acceleration of $1\,g\,(=9.81\ \mathrm{m\,s^{-2}})$ in its own rest frame and the plan is to maintain that acceleration until fly-pass at the Galaxy centre. Captain Kirk is just 35 years old at departure. How old will he be at the fly-pass?

What is the time after the departure of the *Enterprise* at which the commanders on Earth lose all chance of sending a successful light signal that would cancel Captain Kirk's orders and recall him and his starship to Earth?

8.9 An observer O is moving with three-velocity **u** in an inertial frame Σ. A particle is moving in the same inertial frame with three-velocity **v**.

    1   Confirm that the matrix $\mathbf{L}(-\boldsymbol{\beta}_u)$ of equation (8.23) reduces **u** to zero.

    2   What is the speed of the particle as determined in the rest frame of O?

9 Some developments in special relativity

*9.1 Introduction

As the title indicates, we are going to discuss some developments involving special relativity. In previous chapters we have attempted to provide an introduction to the foundations of the subject and have illustrated that introduction with applications. And in fact, we used the application of special relativity to particle kinematics at speeds approaching that of light as a simple, first step to confidence building for the whole subject (Chapters 2 and 3). The present chapter is concerned not with the foundations but with a few of the subjects that have grown from the success of special relativity. We start by building on the use of four-vectors in particle kinematics and in wave physics. This leads naturally to the wave–particle duality which was discovered by de Broglie and which has its origins in relativity. That duality leads to quantum mechanics and we introduce three equations. The first, due to Schrödinger, was motivated by de Broglie's relation but, as we shall show, is non-relativistic. The two remaining are due to Klein and Gordon and to Dirac. It is appropriate to show that both are relativistically covariant. Otherwise we limit the description to some of the early successes of the Dirac equation, in particular in predicting the existence of anti-particles. In the last part of this chapter we return to electromagnetism, a subject which is in the mainstream of relativity. We derive the equations that transform electric and magnetic fields in Lorentz boosts and demonstrate the covariance of Maxwell's equations. We rewrite these equations in tensor notation to show that electromagnetism is constrained by the imperative of relativistic covariance to be as it is observed to be.

In the early chapters we attempted to be easy on the new readers in the business of using vectors (see Table 3.1). In Section 9.7 and on we have assumed that these readers have been pursuing a course in vectors and vector calculus and have, at least, some familiarity with vector notation in electromagnetism.

*9.2 Relativistic kinematics with four-vectors

The four-vector we are going to use is that of energy and momentum: $p = (E, \mathbf{P}c)$ (see equation (8.40)). For conciseness we shall call it the **four-momentum**. We must recall some properties of four-vectors as they apply to the four-momentum. From Section 8.10 we have:

1   The inner product of a four-momentum of a particle with itself is, of course, an invariant, the square of the rest mass energy:

$$p \cdot p \equiv p^2 = E^2 - \mathbf{P} \cdot \mathbf{P}c^2 = (Mc^2)^2. \tag{9.1}$$

(Note that we are using the notational shorthand $p^2 \equiv p \cdot p$.)

2   The inner product of two different four-momenta is an invariant and in any one inertial frame is given in terms of energies and momenta by (equation (8.37))

$$p_1 \cdot p_2 = E_1 E_2 - \mathbf{P}_1 \cdot \mathbf{P}_2 c^2. \tag{9.2}$$

3   The invariance of the inner product of two four-momenta means that such a quantity may be evaluated in any inertial frame that is convenient. We can say that in a more general way: if the formula for a physical quantity in a given inertial frame may be written in terms of rest masses and of inner products of four-momenta then the formula contains only invariant quantities and each may be evaluated in any inertial frame.

4   We know from Section 8.12 that p is the contribution by one particle to the energy and momentum that is conserved in collisions and decays of particles. We shall express that overall conservation in one equation.

Now we can give some examples and useful ideas on using the four-momentum in the relativistic kinematics of particles.

Example 1. Consider the decay a $\rightarrow$ 1 + 2. Find the energy of 1 in the rest frame of 'a' in terms of the masses involved. With obvious notation we have

$$p_a = p_1 + p_2, \tag{9.3}$$

as the single equation expressing energy and momentum conservation. Thus

$$p_a - p_1 = p_2.$$

Square both sides and we have

$$p_a^2 + p_1^2 - 2p_a \cdot p_1 = p_2^2$$

which says

$$(M_a c^2)^2 + (M_1 c^2)^2 - 2(E_a E_1 - \mathbf{P}_a \cdot \mathbf{P}_1 c^2) = (M_2 c^2)^2.$$

This equation is true in all inertial frames and we evaluate it in the rest frame of 'a' in which $p_a \equiv (E_a, \mathbf{P}_a c) = (M_a c^2, \mathbf{0})$ and we obtain

$$(M_a c^2)^2 + (M_1 c^2)^2 - 2E_1 M_a c^2 = (M_2 c^2)^2,$$

or

$$E_1 = (M_a + M_1 - M_2)c^2/2M_a \quad \text{(in the rest frame of a).} \qquad (9.4)$$

This is just what we found in Section 3.3, equation (3.5).

Example 2. Continuing the last example, suppose we need the energy of particle 1 in the frame in which particle 2 is at rest. Square equation (9.3):

$$p_a^2 = p_1^2 + p_2^2 + 2p_1 \cdot p_2$$

or

$$2p_1 \cdot p_2 = p_a^2 - p_1^2 - p_2^2.$$

Evaluate this in the rest frame of 2:

$$2E_1 M_a c^2 = (M_a c^2)^2 - (M_1 c^2)^2 - (M_2 c^2)^2.$$

Then

$$E_1 = (M_a - M_1 - M_2)c^2/2M_1 \quad \text{(in the rest frame of 2).} \qquad (9.5)$$

Equations (9.4) and (9.5) give the energy of 1 in different rest frames as we have emphasised by stating that immediately after each equation.

Example 3. Notice the technique used to find the energy $E_{\alpha,\beta}$ of a particle α in a frame in which particle $\beta(\neq \alpha)$ is at rest. Consider the inner product $p_\alpha \cdot p_\beta$. Evaluate it in the rest frame of β:

$$p_\alpha \cdot p_\beta = E_{\alpha,\beta} M_\beta c^2. \qquad (9.6)$$

Now turn this equation around:

$$E_{\alpha,\beta} = p_\alpha \cdot p_\beta / M_\beta c^2. \qquad (9.7)$$

The right-hand side of equation (9.7) contains only invariants and therefore the particular energy $E_{\alpha,\beta}$ may be found by evaluating that side from four-momenta p_α and p_β, both measured in any convenient inertial frame.

Note that the rest frame of particle β can be replaced by a rest frame not necessarily associated with a single particle. For example, the centre-of-mass of a group of particles. The four-vector p_β must be replaced by a four-momentum representing the energy and momentum of that centre-of-mass considered as a unit. We shall use that technique in Example 7.

Example 4. Compton scattering re-analysed using four-momenta (see Section 3.10). The notation is given by

$$\gamma + e^- \rightarrow \gamma + e^-$$
$$1 \quad 2 \quad\quad 3 \quad 4.$$

The four-momenta are defined by the equation of energy and momentum conservation

$$p_1 + p_2 = p_3 + p_4.$$

Then

$$p_4^2 = (p_1 + p_2 - p_3)^2$$
$$= p_1^2 + p_2^2 + p_3^2 + 2p_1 \cdot p_2 - 2p_2 \cdot p_3 - 2p_3 \cdot p_1.$$

This equation is true in all frames but we choose to evaluate it in the laboratory frame which we define as the one in which the electron with label 2 is at rest:

$$(M_e c^2)^2 = 0 + (M_e c^2)^2 + 0 + 2(E_1 - E_3)M_e c^2 - 2(E_3 E_1 - c^2 P_3 P_1 \cos\theta_{13}),$$

where $M_2 = M_4 = M_e =$ the rest mass of an electron and θ_{13} is the angle of scatter, that is the angle between the direction of the incident photon (1) and the direction of the scattered photon (3), in the laboratory. Using $Pc = E$ for the photons reduces this equation to

$$\frac{1}{E_3} - \frac{1}{E_1} = \frac{1}{M_e c^2}(1 - \cos\theta_{13}). \tag{9.8}$$

This is equation (3.13) which was derived without using four-momenta. However, a careful examination of the method used will show that it is really just the four-momentum method written in long hand. Note that the manipulation of the four-momenta was aimed at getting $\cos\theta_{13}$ in terms of other observables. However, this is a particularly simple case since two of the particles involved are photons and one electron is at rest.

Example 5. The rest mass M of a particle of four-momentum p is given by

$$Mc^2 = \sqrt{p \cdot p}. \tag{9.9}$$

Now, in Sections 3.11 and 3.12 we discussed the idea of the centre-of-mass energy, symbol W, of a group of particles. Suppose four-momentum p is the sum of all the individual four-momenta p_i, $i = 1, 2, 3, \ldots$. Then from equations (3.15) to (3.17) we have that

$$W^2 = p \cdot p = \left(\sum_i p_i\right)^2. \tag{9.10}$$

In Section 3.11 we also stressed the fact that the centre-of-mass energy was equal to the rest mass energy of a particle that if it existed might decay into this group of particles. So we can associate a rest mass M with this group where $M = W/c^2$. Or, alternatively

$$(Mc^2)^2 = p \cdot p = \left(\sum_i p_i \right)^2. \tag{9.11}$$

For two particles colliding

$$W^2 = (p_a + p_b)^2.$$

The most common place to find this used is when an energetic particle 'a' is incident on a target particle 'b' that is stationary in the laboratory. Evaluation in the laboratory frame gives equation (3.21).

Example 6. Missing mass. Consider a particle reaction in which an energetic negatively charged pion (π^-) strikes a stationary proton (p, not to be confused with a four-momentum!):

$$\begin{array}{cccccc} \pi^- & + & p & \rightarrow & X^- & + & p, \\ 1 & & 2 & & 3 & & 4 \end{array}$$

X represents all particles other than the proton that are the products of the collision. (In this reaction X must have net charge of -1 unit where one unit is the charge on the proton.) If we know the energy of the incident pion then a measurement of the direction and energy of the final state proton allows a measurement of the mass of the system X. To see how this works consider the conservation of the total four-momentum:

$$p_1 + p_2 = p_3 + p_4,$$

where

$$p_3 = \sum_i p_i = \text{sum of the four-momenta of all the particles in X.}$$

Now, from equation (9.11) the rest mass of the system X, M_X, is given by

$$(M_X c^2)^2 = p_3 \cdot p_3 \equiv p_3^2.$$

We can find p_3 from

$$p_3 = p_1 + p_2 - p_4.$$

Thus

$$(M_X c^2)^2 = p_3^2$$
$$= p_1^2 + p_2^2 + p_4^2 + 2p_1 \cdot p_2 - 2p_2 \cdot p_4 - 2p_4 \cdot p_1.$$

As usual this may be evaluated in any inertial frame. Evaluating in terms of the energy and momentum components in the laboratory (2 at rest) gives

$$(M_X c^2)^2 = (M_\pi c^2)^2 + 2(M_p c^2)^2 + 2(M_p c^2)(E_\pi - E_4) - 2(E_4 E_\pi) + 2(\mathbf{P}_4 \cdot \mathbf{P}_\pi)c^2.$$

Thus M_X may be found from a knowledge of the masses of pion (M_π) and proton (M_p), and of the momentum three-vectors of the incident pion ($\mathbf{P}_\pi$) and the final proton ($\mathbf{P}_4$). M_X is sometimes called a missing mass: it may be measured without direct observation of any of the group of particles of which X is made up.

Example 7. Finding the angle between two momenta in a given inertial frame. Consider a group of particles 1, 2, 3, ..., n, each identified and with all momenta measured in the laboratory. Suppose we wish to find the angle between the momenta $\mathbf{P}_1$ and $\mathbf{P}_2$ in the rest frame (= centre-of-mass) of all n particles, which we call Σ_{cm}. The four-momentum for this system is

$$p_1 + p_2 + p_3 + \cdots + p_n \equiv p.$$

Then the centre-of-mass energy is $W = \sqrt{p \cdot p}$. By Example 3 above the energy E_i of particle i in this centre-of-mass (Σ_{cm}) is

$$E_i = \frac{p_i \cdot p}{W} = \frac{p_i \cdot p}{\sqrt{p \cdot p}}.$$

(This is correct because we can think of the n particles being a particle of mass W and four-momentum p.) Now consider $p_1 \cdot p_2$ in Σ_{cm}:

$$p_1 \cdot p_2 = E_1 E_2 - \mathbf{P}_1 \cdot \mathbf{P}_2 c^2 = E_1 E_2 - P_1 P_2 c^2 \cos\theta_{12},$$

where θ_{12} is the (required) angle between $\mathbf{P}_1$ and $\mathbf{P}_2$ in Σ_{cm}. Then

$$
\begin{aligned}
\cos\theta_{12} &= \frac{E_1 E_2 - p_1 \cdot p_2}{P_1 P_2 c^2} \\[2mm]
&= \frac{(p_1 \cdot p)(p_2 \cdot p) - (p_1 \cdot p_2)W^2}{W^2 \sqrt{E_1^2 - M_1^2 c^4} \sqrt{E_2^2 - M_2^2 c^4}} \\[2mm]
&= \frac{(p_1 \cdot p)(p_2 \cdot p) - (p_1 \cdot p_2)(p \cdot p)}{\sqrt{(p_1 \cdot p)^2 - p_1^2 p^2} \sqrt{(p_2 \cdot p)^2 - p_2^2 p^2}}.
\end{aligned}
$$

(We have used $p^2 = W^2$, $p_1^2 = (M_1 c^2)^2$, $p_2^2 = (M_2 c^2)^2$.) This looks very complicated but it has a huge virtue: the right-hand side contains only terms that are the inner product of two four-momenta and each of which is therefore a Lorentz invariant. Each term may be calculated in any inertial frame in which the components of the four-vectors are known. From the manner of setting this problem that would be the 'laboratory'. Thus we have found the cosine of an angle in an inertial frame possibly remote from the laboratory without making any Lorentz transformations.

Problems 9.1–9.6 will provide practice in the use of four-vectors in relativistic kinematics.

*9.3   More on waves

In Section 8.10 we discovered the wave four-vector $k = (\omega/c, \mathbf{k})$. Let us look at some of its properties.

a. The plane wave $\exp(ik \cdot x) = \exp i(\omega t - \mathbf{k} \cdot \mathbf{r})$ has a phase $\omega t - \mathbf{k} \cdot \mathbf{r}$ and the wave front moves in the direction of $\mathbf{k}$. If $\mathbf{n}$ is a unit vector in that direction so that $\mathbf{k} = \mathbf{n}|\mathbf{k}|$ we can put

$$\omega\, dt - |\mathbf{k}|\mathbf{n} \cdot d\mathbf{r} = 0. \tag{9.12}$$

Then the speed of the wavefront in the direction of $\mathbf{n}$ is

$$v_{\text{ph}} = \mathbf{n} \cdot \frac{d\mathbf{r}}{dt} = \frac{\omega}{|\mathbf{k}|}. \tag{9.13}$$

This is the so-called **phase velocity**. (Although called a velocity it is, in our convention, a speed. We shall conform to usage and use velocity in this case, and in the case of *group velocity*.)

b. A plane electromagnetic wave in a vacuum has a phase speed of c so that

$$\omega/|\mathbf{k}| = c \tag{9.14}$$

and its wave four-vector is null. ($\omega^2/c^2 - \mathbf{k} \cdot \mathbf{k} = 0$, see Section 8.11.) Therefore in any other inertial frame the phase velocity is

$$\omega'/|\mathbf{k}'| = c,$$

as must be the case.

c. Consider our usual arrangement of Σ' and Σ. A plane electromagnetic wave will have a wave four-vector $k' = (\omega'/c, \mathbf{k}')$ in Σ'. In Σ the same wave will have wave four-vector $k = (\omega/c, \mathbf{k})$. The relation between k and k' is given by a Lorentz transformation:

$$k = \mathbf{L}(\beta_u)k'$$

or

$$\begin{pmatrix} \omega/c \\ k_x \\ k_y \\ k_z \end{pmatrix} = \begin{pmatrix} \gamma_u & 0 & 0 & \gamma_u\beta_u \\ 0 & 1 & 0 & 0 \\ 0 & 0 & 1 & 0 \\ \gamma_u\beta_u & 0 & 0 & \gamma_u \end{pmatrix} \begin{pmatrix} \omega'/c \\ k'_{x'} \\ k'_{y'} \\ k'_{z'} \end{pmatrix} \tag{9.15}$$

$$= \begin{pmatrix} \gamma_u(\omega'/c + \beta_u k'_{z'}) \\ k'_{x'} \\ k'_{y'} \\ \gamma_u(k'_{z'} + \beta_u\omega'/c) \end{pmatrix}. \tag{9.16}$$

d. The aberration of light again: suppose our plane wave is electromagnetic with $\mathbf{k}'$ in the $x'z'$ plane of Σ' at an angle $\theta'(= \arccos(k'_z/|\mathbf{k}'|))$ to the z'-axis. Then from equations (9.15) and (9.16) we have that

$$\omega = \gamma_u(\omega' + \beta_u c|\mathbf{k}'|\cos\theta') = \gamma_u\omega'(1 + \beta_u\cos\theta') \tag{9.17}$$

and if $\mathbf{k}$ is at an angle θ to the z-axis in Σ

$$\cos\theta = \frac{k_z}{|\mathbf{k}|} = \frac{\cos\theta' + \beta_u}{1 + \beta_u\cos\theta'} \tag{9.18}$$

and

$$\tan\theta = \frac{k_x}{k_z} = \frac{\sin\theta'}{\gamma_u(\cos\theta' + \beta_u)}. \tag{9.19}$$

Equations (9.17) and (9.18) are seen to be the same as equations (6.9) and (6.10) if we recall that $f = \omega/2\pi$. The latter equations were derived by treating photons as particles with energy $E = hf$ and momentum $|\mathbf{P}| = E/c$. Equation (9.17) gives the longitudinal relativistic Doppler effect. If $\theta' = 0$ in Σ' the wave is moving in the direction of the z-axis in Σ and then

$$\omega = \gamma_u\omega'(1 + \beta_u) = \omega'\sqrt{\frac{1 + \beta_u}{1 - \beta_u}}.$$

This equation is the same as the result obtained in the particle method (equation (6.14)).

Thus we obtain the same results under Lorentz transformations if we treat electromagnetic waves as waves or as particles. This duality needs further investigation because we might expect the particle nature of light to be associated with the group velocity of an electromagnetic wave rather than with its phase velocity. This matter is resolved in the next section.

e. *Group velocity.* The phase velocity was defined in (a) above (equation (9.13)). A point of maximum displacement of the wave travels with this velocity. However, the plane wave, $\exp(i\mathbf{k}\cdot\mathbf{x})$, is an idealised concept and most waves do not have the plane wave's property of existing for all time. A real wave passing a point will grow in amplitude and then decline. Such a wave does not have a unique frequency but is a superposition of a range of frequencies. It is called a *group* and the position of maximum amplitude moves with the **group velocity**. In elementary wave theory the group velocity v_g is given by

$$v_g = \frac{d\omega}{d|\mathbf{k}|}. \tag{9.20}$$

This equation, plus that for the phase velocity (equation (9.13))

$$v_{ph} = \frac{\omega}{|\mathbf{k}|}$$

make an easy-to-remember pair.

We can now briefly consider a wave motion not moving with the speed of light in a vacuum. The wavefront of Figure 8.7 has a speed $v_{ph} = \omega/|\mathbf{k}|$ (equation (9.13)). Then its wave four-vector may be written

$$k = (\omega/c, \mathbf{k}) = (\omega/c, \mathbf{n}\omega/v_{ph}). \tag{9.21}$$

This means that in the transformation equation (9.15) the components of $\mathbf{k}$ must be replaced with the components of $\mathbf{n}\omega/v_{ph}$, and similarly for k'. The result is that the aberration formula for the direction of the normal to the wave front is different from equations (9.18) and (9.19). For example, in the same circumstances of $\mathbf{k}'$ in the $x'z'$ plane but $v'_{ph} \neq c$, equation (9.19) becomes

$$\tan\theta = \frac{\sin\theta'}{\gamma_u(\cos\theta' + \beta_u v'_{ph}/c)}. \tag{9.22}$$

Equations (9.21) and (9.22) show two things:

1 The case of an electromagnetic wave with $v_{ph} = c$ is a special case.
2 The wavefront velocity $\mathbf{n}v_{ph}$ cannot be incorporated into the space components of a four-velocity.

The reader is invited to apply these ideas in Problems 9.7–9.10.

*9.4 Einstein to de Broglie

In 1905 Einstein proposed that electromagnetic waves were quantised and was thereby able to explain the properties of the photoelectric effect. Later, the name photon was given to this 'light' quantum and now we are accustomed to think of photons of all energies as particles. We discussed this particle–wave connection in Sections 2.3 and 3.9. In 1923, Louis de Broglie extended this idea to all particles and his result is often expressed in terms of a wavelength λ associated with a particle of momentum P through the relation

$$\lambda = h/P, \tag{9.23}$$

where h is Planck's constant. How is this de Broglie result related to special relativity? In Table 9.1 boxes contain all the information required to obtain the relations of particle–wave duality. Box 1 shows the parameters of a wave and, in particular, the components of the wave four-vector k and its relation to the phase at space–time point x. Box 2 reminds us of the kinematic parameters of a moving particle. In Box 3 the duality is quantified by putting

$$\hbar ck = p, \tag{9.24}$$

where $\hbar = h/2\pi$. From equation (9.24) follows

$$E = \hbar\omega \quad \text{and} \quad \mathbf{P} = \hbar\mathbf{k}. \tag{9.25}$$

Table 9.1 The parameters of waves and particles that become related in wave–particle duality. The connection is made (at ∗ in the table) by equating the particle four-momentum (p) to $\hbar c$ times the relativistic wave four-vector (k).

Box 1 Wave parameters (from Section 8.9)

f = frequency
ω = angular frequency = $2\pi f$
λ = wavelength
$\mathbf{k}$ = wave three-vector, $|\mathbf{k}| = 2\pi/\lambda$
$k = (\omega/c, \mathbf{k})$ = wave four-vector
$x = (ct, \mathbf{r})$ = coordinate four-vector
Phase at space–time point $x = k \cdot x = \omega t - \mathbf{k} \cdot \mathbf{r}$

Box 2 Particle parameters

$p = (E, \mathbf{P}c)$ = four-momentum
Mc^2 = rest mass energy = $\sqrt{p \cdot p} = \sqrt{E^2 - \mathbf{P} \cdot \mathbf{P}c^2}$
$\mathbf{v}$ = velocity = $\mathbf{P}c^2/E$
$v = |\mathbf{v}|$ = speed
$P = |\mathbf{P}|$ = momentum

Box 3 Particle–wave duality

We need
h = Planck's constant and $\hbar = h/2\pi$
∗ Assume $p = \hbar c k$
which is $(E, \mathbf{P}c) = \hbar c(\omega/c, \mathbf{k})$
Then $E = \hbar \omega$ and $\mathbf{P} = \hbar \mathbf{k}$
  $\downarrow$ $\downarrow$
  $E = hf$ $\lambda = h/P$
  Einstein 1905   de Broglie 1923

Box 4 Wave velocities

Phase velocity (speed) = $v_{ph} = \omega/|\mathbf{k}|$
From elementary wave theory:
Group velocity (speed) = $v_g = \mathrm{d}\omega/\mathrm{d}|\mathbf{k}|$

Box 5 Particle–wave velocities

$v_{ph} = \omega/|\mathbf{k}| = E/P = c^2/v$

$v_g = \mathrm{d}\omega/\mathrm{d}|\mathbf{k}| = \mathrm{d}E/\mathrm{d}P = Pc^2/E = v$
Then
$v_{ph} v_g = c^2$

and for photons in a vacuum

$v_{ph} = v_g = c$

These are alternative ways of writing Einstein's relation $E = hf$ for the photon and de Broglie's $\lambda = h/P$, for all particles.

Many experiments have been performed that confirm the de Broglie relation for molecular, atomic and sub-atomic particles. The photoelectric effect validated Einstein's relation but only in so far as the fact that the frequency was deduced from the known wavelengths of light. Thus the frequency may be measured directly only for the low frequency end of the electromagnetic spectrum and then only for beams of coherent photons, as in a radio wave. There appear to be no circumstances in which the frequency of a matter wave has any significance for observable physics except through $E = \hbar\omega$. In that case the zero of energy may be of no special relevance. (In Schrödinger's time-independent equation the zero of energy (E in equation (9.28)) occurs when the particle's kinetic energy is zero if the potential is zero. In Dirac's equation (equation (9.32)) for a free particle, the E is the rest mass energy when the momentum is zero. However, if two quantum mechanical states are superimposed but have different energies E_1 and E_2 then the angular frequency difference $\Delta\omega = (E_1 - E_2)/\hbar$ may have observable effects.

We are now in a position to find relations between the velocities involved. For a wave the phase velocity is given by equation (9.13) and in Box 4 of Table 9.1:

$$v_{ph} = \omega/|\mathbf{k}|.$$

Using the Einstein and de Broglie relations this gives for the phase velocity of the wave associated with any particle at any attainable speed

$$v_{ph} = c^2/v, \tag{9.26}$$

where v is the particle's speed. Note that since $v < c$, we have $v_{ph} > c$ and $v_g < c$. The fact that $v_{ph} > c$ causes no problem since no information can be transported except by the group. Box 4 also gives the group velocity of a wave. These results are translated into particle quantities in Box 5. The result is:

$$v_g = v. \tag{9.27}$$

This is just what we would expect and require. The wave group velocity is equal to the particle's speed. We note also the final results: for all particles, including photons

$$v_{ph}v_g = c^2.$$

For photons in a vacuum we have $v_{ph} = v_g = c$.

+9.5   Schrödinger's wave equation

This section and the following are really a story about the beginnings of the profound effect that special relativity had on quantum mechanics.

In 1926, Erwin Schrödinger, building on the de Broglie idea of a wave associated with a particle, proposed a differential equation for a wave function that represented the state of a particle. His notes (Moore[1]) indicate that he started from the classical differential equation satisfied by the function $\psi(\mathbf{r})$ describing the spatial variation of a stationary wave. The equation is:

$$\nabla^2 \psi(\mathbf{r}) + \mathbf{k} \cdot \mathbf{k}\psi(\mathbf{r}) = 0, \tag{9.28}$$

where

$$\nabla^2 \equiv \frac{\partial^2}{\partial x^2} + \frac{\partial^2}{\partial y^2} + \frac{\partial^2}{\partial z^2},$$

and $\mathbf{k}$ is the wave three-vector which satisfies $|\mathbf{k}| = 2\pi/\lambda$. It is at this point that de Broglie contributes. Substituting $\lambda/2\pi = \hbar/P$, where $\hbar$ is Planck's constant divided by 2π, gives

$$\mathbf{k} \cdot \mathbf{k} = P^2/\hbar^2 = 2MT/\hbar^2.$$

In the last equality, M is the mass and T is the kinetic energy of the particle. This is a non-relativistic relation. The kinetic energy is the total (non-relativistic) energy E less the potential energy V. Thus equation (9.28) becomes

$$\hbar^2 \nabla^2 \psi(\mathbf{r}) + 2M(E - V(\mathbf{r}))\psi(\mathbf{r}) = 0, \tag{9.29}$$

where the potential energy varies with the position, and $\psi(\mathbf{r})$ is the wave function. This is Schrödinger's time-independent wave equation. He solved it for the hydrogen atom ($V(r) = -e^2/(4\pi\varepsilon_0 r)$), obtaining the same energy levels as are found using the simple Bohr model.

The reader might ask what is the connection with special relativity. The de Broglie relation has its basis in relativity and Schrödinger obtained and solved a supposedly relativistic form of equation (9.28), but abandoned it in favour of the non-relativistic form as it failed to give results in agreement with observation. The modern approach is that the energy and momentum are represented by differential operators acting on the wave function. Suppose we have a free particle of total energy E and momentum P and we assume it is represented by a plane wave function $\psi(t, \mathbf{r}) = \exp(-i[\omega t - \mathbf{k} \cdot \mathbf{r}])$. Then using the de Broglie relations (equation (9.25))

$$i\hbar \frac{\partial}{\partial t} \psi(t, \mathbf{r}) = \hbar\,\omega\psi(t, \mathbf{r}) = E\psi(t, \mathbf{r})$$

and

$$-i\hbar \frac{\partial}{\partial x} \psi(t, \mathbf{r}) = \hbar\,k_x\psi(t, \mathbf{r}) = P_x\psi(t, \mathbf{r}),$$

1 Moore, W., *Schrödinger, Life and thought,* Cambridge: Cambridge University Press, 1989.

and similarly for the y- and z-components of the momentum. These equations have the structure:

Operator acting on wave function = observable times the wave function.

The operator–observable connection proposed is therefore

$$E \to i\hbar \frac{\partial}{\partial t}, \quad P_x \to -i\hbar \frac{\partial}{\partial x}, \quad P_y \to -i\hbar \frac{\partial}{\partial y}, \quad P_z \to -i\hbar \frac{\partial}{\partial z}. \qquad (9.30)$$

To obtain Schrödinger's equation we apply this to the non-relativistic expressions

$$T + V(\mathbf{r}) = E \quad \text{where } T = P^2/2M.$$

This gives

$$-\frac{\hbar^2}{2M} \nabla^2 \psi(t, \mathbf{r}) + V(\mathbf{r})\psi(t, \mathbf{r}) = i\hbar \frac{\partial}{\partial t} \psi(t, \mathbf{r}). \qquad (9.31)$$

This is Schrödinger's time-dependent wave equation. For stationary particle states (for example, the bound states of an electron in the hydrogen atom) where the energy is constant, the right-hand side may be replaced by $E\psi(t, \mathbf{r})$ which gives equation (9.29).

The space and time coordinates do not appear in an equivalent way in Schrödinger's equations. Therefore the latter can not be relativistically covariant. If we maintain the operator–observable connection of equation (9.30) then we can make a wave equation from

$$E^2 - P^2 c^2 = (Mc^2)^2,$$

where E is now the relativistic total energy. The equation is

$$-(\hbar c)^2 \frac{\partial^2}{\partial (ct)^2} \psi(t, \mathbf{r}) + (\hbar c)^2 \nabla^2 \psi(t, \mathbf{r}) = (Mc^2)^2 \psi(t, \mathbf{r}). \qquad (9.32)$$

This is the Klein–Gordon equation for a free particle. The differential operator that appears on the left-hand side of the equation is

$$\nabla^2 - \frac{\partial^2}{\partial (ct)^2}$$

and is called the d'Alembertian operator. It is Lorentz covariant (see Section 9.11). This means that equation (9.32) is true in all inertial frames provided that $\psi(t, \mathbf{r})$ is Lorentz invariant, that is a scalar. (Under a Lorentz transformation $(ct, \mathbf{r}) \to (ct', \mathbf{r}')$, the function $\psi(t, \mathbf{r}) \to \psi'(t', \mathbf{r}') = \psi(t, \mathbf{r})$.) Such a single-component, scalar wave function is interpreted as representing a particle without spin. The Klein–Gordon equation was proposed in 1926 following Schrödinger's first publications on his wave equation. However, it was initially rejected due to unacceptable faults:

1   It predicted the existence of solutions with negative energy.
2   It allowed negative values for the probability of finding the particle per unit volume of space (probability density).

It was some years before the Klein–Gordon equation was to be rescued and given a proper interpretation. Before then, another covariant wave equation was discovered by Dirac.

+9.6   The Dirac equation

P. A. M. Dirac set out to find a first-order linear differential equation for a wave function that would describe the behaviour of a single free particle, consistent with the principles of special relativity, but without the faults of the Klein–Gordon equation. To avoid negative probabilities he knew that the differential equation would have to be first order in $\partial/\partial t$ and therefore also first order in $\partial/\partial x, \partial/\partial y, \partial/\partial z$. Without that property Lorentz covariance would be impossible. The equation he proposed was

$$i\hbar \frac{\partial}{\partial ct}\psi(ct, \mathbf{r}) = -i\hbar \left(\alpha_x \frac{\partial}{\partial x} + \alpha_y \frac{\partial}{\partial y} + \alpha_z \frac{\partial}{\partial z} \right) \psi(ct, \mathbf{r}) + \beta Mc\psi(ct, \mathbf{r})$$

$$E/c \qquad\qquad\qquad P_x \quad\quad P_y \quad P_z \qquad\qquad\qquad Mc \qquad (9.33)$$

Below the equation we have put the kinematic quantities related to the differential operators appearing in the various terms. Note that Dirac used the operators that also give the Schrödinger equations. First it is clear that the quantities $\alpha = (\alpha_x, \alpha_y, \alpha_z)$, and β cannot be real numbers since P and Mc do not add linearly to make E/c.

Dirac required that the wave function $\psi(ct, \mathbf{r})$ satisfy the Klein–Gordon equation. The procedure is to square the operators on both sides of equation (9.33). This gives the Klein–Gordon equation if the quantities $\alpha_x, \alpha_y, \alpha_z, \beta$ satisfy certain relations; these are:

$$a_i^2 = \beta^2 = \mathbf{I}, \quad \alpha_i\beta + \beta\alpha_i = 0, \qquad i = x, y, z,$$
$$\alpha_i\alpha_j + \alpha_j\alpha_i = 0, \quad i, j = x, y, z, \quad i \neq j.$$

The simplest representation of these quantities is by 4×4 matrices and $\mathbf{I}$ is then the 4×4 unit matrix. That means $\psi(ct, \mathbf{r})$ is a four-element column matrix. For a particle of a given momentum there are four solutions (each a column matrix with four elements). All solutions have positive probability densities, but two have negative energy. Thus Dirac had removed only one of the Klein–Gordon faults. We return to that problem shortly.

The question of the proof of Lorentz covariance of the equation remains. We sketch how this is done. First we make an improvement in notation. We replace the four-vector written as $(ct, \mathbf{r})$ by $x = x^0, x^1, x^2, x^3 = x^\mu$ (Section 8.10) and put $\gamma^0 = \beta$, $\gamma^i = \beta\alpha_i$, $i = 1, 2, 3$, with the obvious change in index notation, that is $\alpha_1 = \alpha_x$, and so on. Multiply equation (9.33) by β and we find

$$i\hbar \left(\sum_{\mu=0}^{3} \gamma^\mu \frac{\partial}{\partial x^\mu} \right) \psi(x) - Mc\psi(x) = 0. \tag{9.34}$$

We can imagine an observer O in Σ of our standard arrangement of inertial reference frames setting up this equation. By the principle of relativity a second observer O′ in Σ' will set up an equation identical in form and therefore identical in physical content:

$$i\hbar \left(\sum_{\mu=0}^{3} \bar{\gamma}^\mu \frac{\partial}{\partial x'^\mu} \right) \psi'(x') - Mc\psi'(x') = 0, \tag{9.35}$$

where $x' = \mathbf{L}(-\beta_u)x$, the normal transformation of coordinates from Σ to Σ' (Section 7.4 and equations (7.4) and (7.5)). The $\bar{\gamma}$ matrices must satisfy the same relations as do the $\bar{\gamma}$ matrices in equation (9.34); these in turn depend upon the the relations satisfied by the matrices α and β. There are in fact many choices for α and β, all equivalent in their effect, so we may allow the four $\bar{\gamma}$ matrices to be the same as the four γ matrices.

We are assured of covariance if

1   there is an algorithm enabling O′ to calculate $\psi'(x')$ from $\psi(x)$, and
2   $\psi'(x')$ satisfies equation (9.35) and thereby describes to O′ the transformed $(P \rightarrow P', E \rightarrow E', \ldots)$ particle state that $\psi(x)$ describes to O.

So the problem is to find a 4×4 matrix S with the property that acting on the four-component column matrix $\psi(x)$ gives $\psi'(x')$. S will be a function of the parameters of the transformation from Σ to Σ'. (It is worth noting that although $\psi(x)$ is represented by a four-element column matrix it is not a four-vector and therefore does not transform as does x. The name given to such an object is *spinor*.) Substituting $S\psi(x) = \psi'(x')$ into equation (9.34) and making the necessary transformations of the differential operators from Σ to Σ' (see Table 9.2) gives an equation which must be the same as equation (9.35) for covariance. That gives an equation for S. That equation can be solved and thus the algorithm exists and the Dirac equation is covariant. The details of all this are complicated but may be found in some text books on quantum mechanics, for example, in that by Bjorken and Drell.[2]

Dirac's equation is a direct outcome of the marriage of special relativity and quantum mechanics. Although it is not appropriate for us to discuss the details of

2 Bjorken, J. D. and Drell, S. D., *Relativistic quantum mechanics*, New York: McGraw-Hill Inc., 1964.

the development of this equation, it is right to mention some of its initial successes since these depend on the inheritance from relativity.

1   The two solutions having positive energy correspond to the two degrees of freedom that an intrinsic spin angular momentum of $\hbar/2$ has in quantum mechanics. In the jargon of the subject, the Dirac equation describes the behaviour of a spin-$\frac{1}{2}$ particle.

2   In 1928, the electron was known to have spin $\frac{1}{2}$. Applied to this particle, the Dirac equation predicted that it had a magnetic dipole moment equal to $g(e/2M_e c)(\hbar/2)$ with the gyromagnetic ratio $g = 2$ (a pure number). That magnetic moment, in atomic units, is one Bohr magneton, consistent with experimental results on the anomalous Zeeman effect. (M_e is the mass of the electron.)

3   The Dirac equation can be solved for the case of the hydrogen atom. The solution gave the correct energy levels including the fine structure (absent from the Schrödinger solution). This was an interesting result as previously the predictions for the fine structure effects based on the existence of electron spin had been wrong by a factor of 2, corrected only by including the effect of the Thomas precession (see Section 8.13).

There were some difficulties not immediately apparent at the time. The gyromagnetic ratio is not precisely 2. The Dirac solution for the hydrogen atom predicted that the $2S_{1/2}$ and $2P_{1/2}$ levels, for example, had the same energy. They differ by a very small amount that was not measured accurately until 1947. There were also corrections for the hyperfine splitting due to the proton spin but these were understood and not directly of concern to testing this relativistic quantum theory.

However, more urgent was an understanding of the negative energy solutions. In 1930, Dirac proposed a model in which the vacuum contained a *sea* of negative energy states, starting at $-M_e c^2$ and extending to negative infinite energy, filled with electrons according to the Pauli exclusion principle (only one electron occupying each state). This bottomless sea of negative energy electrons would be unobservable. However, the absence of an electron from one of these states would be observed to behave like a positively charged particle of spin $\frac{1}{2}$ with positive energy. Dirac attempted to find how such a 'hole' in the sea could describe a proton, the only other then known positive massive elementary particle, but did not succeed. He came round to the symmetrical view that the electrically positive holes would have the same mass as the electron. Thus this was a prediction for the existence of an unknown particle. In 1933 a particle with these properties was observed in cosmic radiation and later named the positron. The positron is the anti-particle to the electron (Section 2.5). We now know that it is not just the electron that has an anti-particle. All charged sub-atomic particles, and some uncharged, have distinct antiparticles. Dirac's model of the vacuum has been replaced by a proper interpretation of the negative energy solutions both of his equation and of the Klein–Gordon equation. That interpretation prescribes how to identify these solutions as describing positive energy anti-particles and allows the use of those solutions in calculations about processes involving anti-particles. And the problems of the exact value of the electron magnetic moment and of the hydrogen $2S_{1/2}$–$2P_{1/2}$ level splitting have been solved.

The last paragraph serves to finish our summary of some of the first results that the Dirac equation produced. However, the prediction and discovery of the anti-particle to the electron was a momentous prelude to developments and discoveries in sub-atomic physics, both experimental and theoretical, that have continued since 1933 and in which the Dirac equation has played a major role. Thus the fruitfulness of his equation places Dirac with Schrödinger among the giants of physics. An up-to-date and accessible account of theoretical developments in the field of sub-atomic physics is to be found in the book by Aitchison and Hey.[3] Interestingly, it starts where we have stopped. Their first chapter is entitled 'The Klein–Gordon and Dirac Equations, and the Interpretation of Their Negative Energy Solutions'.

Before leaving this section there are two points to be made.

1   Non-relativistic Schrödinger was not eclipsed by relativistic Dirac. The former's equations have proved to be invaluable in the theory of phenomena not involving relativistic speeds. For example, the theory of chemical structures and reactions depends primarily on Schrödinger's equations applied to the electronic structure of atoms and molecules. However, recent calculations have used approximations based on Dirac, particularly for the electronic structure of heavy elements. Dirac was essential to understanding sub-atomic physics. But both can trace their beginnings to special relativity.

2   In this and the previous two sections we have traced a line of development that went from Einstein to de Broglie to Schrödinger to Dirac. This has not done justice to many others who contributed greatly to this development of quantum mechanics and to the interpretation of its results. Readers who wish to know more about the early years of the subject and about the people involved are referred to the book by Pais.[4]

9.7   Maxwell's equations

Electromagnetism was the outcome of the unification of electricity and magnetism achieved by Maxwell in the middle of the nineteenth century. It was and remains a very successful theory yet it is not covariant under Galilean transformations and in the view of nineteenth century physicists it required the existence of the ether (Section 5.4). This ether remained undetected (see Sections 10.2 and 10.3). All this was the prelude to Einstein's discovery of special relativity. His first postulate demands that Maxwell's equations be covariant under Lorentz transformations. That allowed Einstein to derive the transformations for the electric and magnetic fields. The ether as previously envisaged withered away. Now we recognise that

3 Aitchison, I. J. R. and Hey, A. J. G., *Gauge theories in physics*, 2nd edn, Bristol: Adam Hilger Ltd, 1989.
4 Pais, A., *Inward bound*, Oxford: Oxford University Press, 1989.

electromagnetism is a relativistic theory and a paradigm for all theories designed to describe any interaction between bodies.

It is sometimes stated that Maxwell's equations may be derived from Coulomb's law and special relativity alone. Jackson[5] in his book entitled *Classical electrodynamics* shows that this is not the case and that some additional assumptions are required. A proper treatment is given by Schwartz.[6]

In the following sections we shall derive the equations that transform the electric field **E** and magnetic induction **B** between inertial frames (Section 9.8), review how Einstein used the covariance of Maxwell's equations (Section 9.9), and introduce the electromagnetic four-vector potential (Section 9.10). This programme only lightly touches what is, in fact, a vast subject.

At this point we quote Maxwell's equations for fields and specified sources in a vacuum. We shall not discuss the macroscopic equations that must apply in the presence of matter in addition to sources. In conventional notation and using SI units we have:

$$\text{div } \mathbf{E} = \rho/\varepsilon_0, \qquad \text{div } \mathbf{B} = 0,$$

$$\text{curl } \mathbf{E} = -\frac{\partial \mathbf{B}}{\partial t}, \qquad \text{curl } \mathbf{B} = \mu_0 \left(\mathbf{j} + \varepsilon_0 \frac{\partial \mathbf{E}}{\partial t} \right), \tag{9.36}$$

$$\text{div } \mathbf{j} + \frac{\partial \rho}{\partial t} = 0,$$

where ε_0 is the permittivity and μ_0 is the permeability of the vacuum, ρ is the charge density and $\mathbf{j}$ is the current density. These equations predict the existence of electromagnetic waves moving with a speed $1/\sqrt{\varepsilon_0 \mu_0}$ in a vacuum. Maxwell showed that this quantity was close to the then measured speed of light and thereby identified light as an electromagnetic phenomenon (Section 5.4).

Covariance of Maxwell's equations requires that if equations (9.36) are correct in inertial frame Σ then in the inertial frame Σ' they have the same form. Thus, at a given space–time point the transformation from Σ to Σ' causes

$$(ct, \mathbf{r}) \rightarrow (ct', \mathbf{r}'), \quad \mathbf{E} \rightarrow \mathbf{E}', \quad \mathbf{B} \rightarrow \mathbf{B}', \quad \rho \rightarrow \rho', \quad \mathbf{j} \rightarrow \mathbf{j}'$$

and

$$\text{div } \mathbf{E} = \rho/\varepsilon_0 \rightarrow \text{div}' \mathbf{E}' = \rho'/\varepsilon_0, \quad \text{curl } \mathbf{E} = -\frac{\partial \mathbf{B}}{\partial t} \rightarrow \text{curl}' \mathbf{E}' = -\frac{\partial \mathbf{B}'}{\partial t'},$$

and similarly for the divergences and curls of **B** and **B'** and for the continuity equation. The primed differential operators are to be with respect to the space–time coordinates in Σ'. Note that the permittivity and permeability are separately the same in each frame. This is essential if the equations are to be consistent with the relativity principle (for example, Coulomb's law must be the same) and the invariance of the speed of light.

5 Jackson, J. D., *Classical electrodynamics*, 2nd edn, New York: John Wiley & Sons, 1975, pp. 578–81.
6 Schwartz, M., *Principles of electrodynamics*, New York: McGraw-Hill Inc., 1972, Chapter 3.

9.8   The Lorentz transformation of E and B

In this section we give a very simple way of finding the equations of the Lorentz transformation of the electric field and magnetic induction. We suppose that we have decided to define the fields **E** and **B** by their effect on a moving particle, charge q, velocity **v**. Then the relevant equation is (equation 5.20)

$$\frac{d\mathbf{P}}{dt} = q(\mathbf{E} + \mathbf{v} \times \mathbf{B}).$$ (9.37)

So by measuring the rate of change of momentum of the particle of charge q it is possible to measure the fields **E** and **B**. However this equation is not naturally covariant. We can correct this as follows:

We put $dt = \gamma_v d\tau$, where τ is the proper time for the particle and is an invariant. As usual $\gamma_v = 1/\sqrt{1 - v^2/c^2}$ (see Section 7.13 for a reminder about proper time). Put $\mathbf{v} = \boldsymbol{\beta}_v c$. Then we have

$$\frac{d\mathbf{P}}{d\tau} = q(\gamma_v \mathbf{E} + c\gamma_v \boldsymbol{\beta}_v \times \mathbf{B}).$$ (9.38)

This equation concerns a three-vector so we need to find an equation for the energy component of the four-momentum for the particle. The rate of energy increase is given by the scalar product of force and velocity. The force is the right-hand side of equation (9.37). Therefore we have

$$\frac{dE}{dt} = q(\mathbf{E} + \mathbf{v} \times \mathbf{B}) \cdot \mathbf{v} = q\mathbf{E} \cdot \mathbf{v}.$$

since $(\mathbf{v} \times \mathbf{B}) \cdot \mathbf{v} = 0$. E is the relativistic particle energy, not the magnitude of **E**. Then the rate of change of E with proper time is

$$\frac{dE}{d\tau} = c\, q\gamma_v \boldsymbol{\beta}_v \cdot \mathbf{E}.$$ (9.39)

By putting the four-momentum $p = (E, \mathbf{P}c)$ we can write equations (9.38) and (9.39) in one matrix equation:

$$\frac{dp}{d\tau} = \frac{d}{d\tau} \begin{pmatrix} E \\ P_x c \\ P_y c \\ P_z c \end{pmatrix} = cq \begin{pmatrix} 0 & E_x & E_y & E_z \\ E_x & 0 & cB_z & -cB_y \\ E_y & -cB_z & 0 & cB_x \\ E_z & cB_y & -cB_x & 0 \end{pmatrix} \begin{pmatrix} \gamma_v \\ \gamma_v\, \beta_{vx} \\ \gamma_v\, \beta_{vy} \\ \gamma_v\, \beta_{vz} \end{pmatrix}$$

Proper time rate of change of four-momentum   Matrix **F**   Particle's fractional four-velocity $\bar{v}/c$

= four-force acting on particle.   (9.40)

Or, more simply

$$\frac{dp}{d\tau} = cq\mathbf{F}\bar{v}/c.$$

The four-vectors in this equation are $dp/d\tau$ and $\bar{v}/c = (\gamma_v, \gamma_v\beta_v)$. Now in transforming a four-vector from to Σ to Σ' we must use the matrix $\mathbf{L}(-\beta_u)$ from Sections 6.2 and 7.4:

$$\frac{dp}{d\tau} \rightarrow \frac{dp'}{d\tau} = \mathbf{L}(-\beta_u)\frac{dp}{d\tau}, \quad \bar{v} \rightarrow \bar{v}' = \mathbf{L}(-\beta_u)\bar{v}.$$

We are assuming charge is invariant. If equation (9.40) is to be covariant, $\mathbf{F}\bar{v}$ must also be a four-vector so that

$$\mathbf{F}\bar{v} \rightarrow \mathbf{F}'\bar{v}' = \mathbf{L}(-\beta_u)\mathbf{F}\bar{v}$$

and in Σ'

$$\frac{dP'}{d\tau} = cq\mathbf{F}'\bar{v}'/c.$$

Now the matrix $\mathbf{L}$ satisfies (Section 6.2)

$$\mathbf{L}(+\beta_u)\mathbf{L}(-\beta_u) = \mathbf{I},$$

where $\mathbf{I}$ is the 4×4 unit matrix, so we can write

$$\mathbf{F}'\bar{v}' = \mathbf{L}(-\beta_u)\mathbf{F}\bar{v} = \mathbf{L}(-\beta_u)\mathbf{F}\,\mathbf{L}(+\beta_u)\mathbf{L}(-\beta_u)\bar{v}$$
$$= \mathbf{L}(-\beta_u)\mathbf{F}\,\mathbf{L}(+\beta_u)\bar{v}'.$$

Therefore

$$\mathbf{F}' = \mathbf{L}(-\beta_u)\,\mathbf{F}\,\mathbf{L}\,(+\beta_u). \tag{9.41}$$

$\mathbf{F}'$ must have the same structure with respect to $\mathbf{E}'$ and $\mathbf{B}'$ as does $\mathbf{F}$ with respect to $\mathbf{E}$ and $\mathbf{B}$ so we put

$$\mathbf{F}' = \begin{pmatrix} 0 & E'_{x'} & E'_{y'} & E'_{z'} \\ E'_{x'} & 0 & cB'_{z'} & -cB'_{y'} \\ E'_{y'} & -cB'_{z'} & 0 & cB'_{x'} \\ E'_{z'} & cB'_{y'} & -cB'_{x'} & 0 \end{pmatrix}. \tag{9.42}$$

Then evaluating the right-hand side of equation (9.41), using $\mathbf{L}(\pm\beta_u)$ from equation (7.5) or (6.5), and $\mathbf{F}$ from equation (9.40) gives

$$\mathbf{F}' = \begin{pmatrix} 0 & \gamma_u(E_x - \beta_u c B_y) & \gamma_u(E_y + \beta_u c B) & E_z \\ \gamma_u(E_x - \beta_u c B_y) & 0 & cB_z & -\gamma_u(cB_y - \beta_u E_x) \\ \gamma_u(E_y + \beta_u c B_x) & -cB_z & 0 & \gamma_u(cB_x + \beta_u E_y) \\ E_z & \gamma_u(cB_y - \beta_u E_x) & -\gamma_u(cB_x + \beta_u E_y) & 0 \end{pmatrix}. \tag{9.43}$$

We may read equations (9.42) and (9.43), element by element, and from the two expressions for $\mathbf{F}'$ find the individual transformation equations. For the transformation from Σ to Σ':

$$E_x \rightarrow E'_{x'} = \gamma_u(E_x - \beta_u c B_y), \qquad c B_{x'} \rightarrow c B'_{x'} = \gamma_u(c B_x + \beta_u E_y),$$

$$E_y \rightarrow E'_{y'} = \gamma_u(E_y + \beta_u c B_x), \qquad c B_{y'} \rightarrow c B'_{y'} = \gamma_u(c B_y - \beta_u E_x),$$

$$\tag{9.44}$$

$$E_z \rightarrow E'_{z'} = E_z, \qquad\qquad\qquad c B_{z'} \rightarrow c B'_{z'} = c B_z.$$

There are some comments that can be made about these results.

1   The inverse transform Σ' to Σ may be formed by interchanging primed and unprimed vector components, and changing the sign of all the β_u factors.
2   We can call the direction of the transformation (along z) the longitudinal direction. Directions perpendicular to longitudinal we call transverse. Then the longitudinal components of $\mathbf{E}$ and $\mathbf{B}$ are unchanged in the transformation. It is worth examining the structure of the transformation of transverse fields for the pattern of the contributions.
3   Remember that these relations apply to fields at a point in a vacuum. The sources of these fields are not at this point.
4   Matrices $\mathbf{F}$ and $\mathbf{F}'$ may appear to give a simple way of dealing with the Lorentz force and the transformations of $\mathbf{E}$ and $\mathbf{B}$. But heed a warning! They do not fit into the universal, rigorous way of doing business in relativity. That is provided by the formalism of tensors. Then $\mathbf{F}$ is replaced by a rank-2 tensor sometimes called the **Faraday**. And equation (9.40) changes from a matrix multiplication on one side to operations involving tensors. The answers are the same. However, tensors allow the representation of physical quantities requiring more indices than the two (row and column number) that a matrix can handle. The same applies to four-vectors: we have represented them by column matrices but they should be thought of as a rank-1 tensors. Some more on this in Section 9.11.
5   This derivation does not show that Maxwell's equations are covariant. That can be done by the methods used in the next section although there we start with assuming that the equations are covariant and sketch the derivation of the rules we produced in equations (9.44).

There is a difficulty with this derivation of the transformation equations for $\mathbf{E}$ and $\mathbf{B}$. Einstein derived these equations (9.44) from the principles of relativity and Maxwell's equations. From that point he proved his version of equation (5.18) which, using Planck's definition of momentum (equation (5.19)), becomes our equation (9.38). Thus it appears as if we are working backwards to derive something used to prove our starting point. However, we wish to put forward the view that the system Maxwell plus Einstein plus Planck is an interlocked structure and that we are free to examine any linkage that we may choose. In elementary electromagnetism $\mathbf{E}$ is defined by equation (4.1) and and we could include

Table 9.2 Lorentz transformation of differential operators

We need to find how to transform between inertial frames the operators such as

$$\frac{\partial}{\partial ct}, \quad \frac{\partial}{\partial x}, \quad \text{and so on.}$$

As usual the coordinates $x = (ct, \mathbf{r})$ are those in inertial frame Σ and $x' = (ct', \mathbf{r}')$ are the same in Σ'. By the chain rule (Lyons,[7] Boas[8]) we can construct the matrix equation

$$
\begin{pmatrix} \dfrac{\partial}{\partial ct} \\[2mm] \dfrac{\partial}{\partial x} \\[2mm] \dfrac{\partial}{\partial y} \\[2mm] \dfrac{\partial}{\partial z} \end{pmatrix}
=
\begin{pmatrix}
\dfrac{\partial ct'}{\partial ct} & \dfrac{\partial x'}{\partial ct} & \dfrac{\partial y'}{\partial ct} & \dfrac{\partial z'}{\partial ct} \\[2mm]
\dfrac{\partial ct'}{\partial x} & \dfrac{\partial x'}{\partial x} & \dfrac{\partial y'}{\partial x} & \dfrac{\partial z'}{\partial x} \\[2mm]
\dfrac{\partial ct'}{\partial y} & \dfrac{\partial x'}{\partial y} & \dfrac{\partial y'}{\partial y} & \dfrac{\partial z'}{\partial y} \\[2mm]
\dfrac{\partial ct'}{\partial z} & \dfrac{\partial x'}{\partial z} & \dfrac{\partial y'}{\partial z} & \dfrac{\partial z'}{\partial z}
\end{pmatrix}
\begin{pmatrix} \dfrac{\partial}{\partial ct'} \\[2mm] \dfrac{\partial}{\partial x'} \\[2mm] \dfrac{\partial}{\partial y'} \\[2mm] \dfrac{\partial}{\partial z'} \end{pmatrix}.
\tag{9.45}
$$

We can now fill in the elements of the square matrix in equation (9.45) for the standard arrangement of inertial reference frames Σ and Σ' (the latter moving with speed u along the z direction in Σ). In particular we have (equation (7.5))

$$
\begin{pmatrix} ct' \\ x' \\ y' \\ z' \end{pmatrix} = \mathbf{L}(-\beta_u) \begin{pmatrix} ct \\ x \\ y \\ z \end{pmatrix} =
\begin{pmatrix}
\gamma_u & 0 & 0 & -\gamma_u \beta_u \\
0 & 1 & 0 & 0 \\
0 & 0 & 1 & 0 \\
-\gamma_u \beta_u & 0 & 0 & \gamma_u
\end{pmatrix}
\begin{pmatrix} ct \\ x \\ y \\ z \end{pmatrix}.
\tag{9.46}
$$

An examination of the differentials required for the square matrix in equation (9.45) shows it to be equal to those in equation (9.46). Note that equation (9.45) transforms the differential operators from Σ' to Σ whereas equation (9.46) transforms the coordinates from Σ to Σ'. Considering one specific direction, namely the transformations from Σ' to Σ, we have for the coordinates

$$x = \mathbf{L}(+\beta_u)\, x'
\tag{9.47}$$

whereas for the differential operators we have

$$\frac{\partial}{\partial x} = \mathbf{L}(-\beta_u)\frac{\partial}{\partial x'}.
\tag{9.48}$$

In equations (9.47) and (9.48) the column matrices of equations (9.45) and (9.46) are represented by $x, x', \partial/\partial x$ and $\partial/\partial x'$. The coordinates $x(x')$ are four-vectors, but the four-component quantities $\partial/\partial x$ and $\partial/\partial x'$ do not transform in the same way and are not, therefore, four-vectors as we have defined them. More on this in Section 9.11.

Since we need the contents of equation (9.48) in Section 9.9 we write it in full:

$$
\begin{pmatrix} \dfrac{\partial}{\partial ct} \\[2mm] \dfrac{\partial}{\partial x} \\[2mm] \dfrac{\partial}{\partial y} \\[2mm] \dfrac{\partial}{\partial z} \end{pmatrix}
=
\begin{pmatrix}
\gamma_u & 0 & 0 & -\gamma_u \beta_u \\
0 & 1 & 0 & 0 \\
0 & 0 & 1 & 0 \\
-\gamma_u \beta_u & 0 & 0 & \gamma_u
\end{pmatrix}
\begin{pmatrix} \dfrac{\partial}{\partial ct'} \\[2mm] \dfrac{\partial}{\partial x'} \\[2mm] \dfrac{\partial}{\partial y'} \\[2mm] \dfrac{\partial}{\partial z'} \end{pmatrix}.
\tag{9.49}
$$

7 Lyons, L., *All you wanted to know about mathematics but were afraid to ask.*, vol. 1, Cambridge: Cambridge University Press, 1995, p. 216.
8 Boas, M. L., *Mathematical methods in the physical sciences*, 2nd edn, New York: John Wiley & Sons, 1983, pp. 166–7.

B in that procedure by using equation (4.3). Once that is done, equations (9.44) follow from the principles of relativity and assuming that $(E, \mathbf{P}c)$ is a four-vector, without mentioning Maxwell's equations.

Problems 9.11–9.13 will give the reader practice in using the transformations of equations (9.44).

+9.9   The covariance of Maxwell's equations

As we have already indicated, Einstein found the transformation equations for **E** and **B** by requiring that Maxwell's equations satisfied the principle of relativity, that is that they were covariant. We shall outline how this was done, thereby confirming this covariance for ourselves. It is straightforward but does require some juggling with partial differentials. We do it first for fields in a vacuum and in the absence of sources at the point where the fields exist.

Let us look at the z component of curl **E**:

$$\frac{\partial E_y}{\partial x} - \frac{\partial E_x}{\partial y} = -\frac{\partial c B_z}{\partial ct}.$$

Replace each differential operator by its expression in terms of differential operators in Σ' as shown in Table 9.2 (equation (9.49)):

$$\frac{\partial E_y}{\partial x'} - \frac{\partial E_x}{\partial y'} = -\gamma \left(\frac{\partial}{\partial ct'} - \beta \frac{\partial}{\partial z'} \right) c B_z.$$

(For conciseness we have dropped the subscripts u from γ_u and β_u of Table 9.2.) Rearranging gives

$$\frac{\partial E_y}{\partial x'} - \frac{\partial E_x}{\partial y'} - \gamma \beta \frac{\partial c B_z}{\partial z'} = -\gamma \frac{\partial c B_z}{\partial ct'}. \tag{9.50}$$

To make progress we must use div **B** $= 0$. Again replacing the differential operators gives

$$\frac{\partial c B_x}{\partial x'} + \frac{\partial c B_y}{\partial y'} + \gamma \left(\frac{\partial}{\partial z'} - \beta \frac{\partial}{\partial ct'} \right) c B_z = 0.$$

Therefore

$$-\gamma \frac{\partial c B_z}{\partial z'} = \frac{\partial c B_x}{\partial x'} + \frac{\partial c B_y}{\partial y'} - \gamma \beta \frac{\partial c B_z}{\partial ct'}. \tag{9.51}$$

Substitute equation (9.51) into (9.50), multiply by γ, use $\gamma^2(1 - \beta^2) = 1$, and rearrange terms. That gives

$$\frac{\partial}{\partial x'} \left[\gamma (E_y + \beta c B_x) \right] - \frac{\partial}{\partial y'} \left[\gamma (E_x - \beta c B_y) \right] = -\frac{\partial c B_z}{\partial ct'}.$$

For covariance this must be identical to $(\text{curl}' \, \mathbf{E}')_z = -\partial c B'_z / \partial ct'$. That is the case if

$$B'_{z'} = B_z, \quad E'_{y'} = \gamma (E_y + \beta c B_x), \quad E'_{x'} = \gamma (E_x - \beta c B_y). \tag{9.52}$$

These are just the same transformations for these components as we derived using the matrix $\mathbf{F}$ in the last section. The procedure of this section can be continued for all parts of Maxwell's source-free equations to show that these transformations are consistent with covariance for these equations.

For the case of fields with sources, that is $\rho \neq 0$ and/or $\mathbf{j} \neq 0$ we again go through the process of changing the differential operators. Start in Σ with

$$(\text{curl} \, c\mathbf{B})_z = \mu_0 c j_z + \frac{\partial E_z}{\partial ct} \tag{9.53}$$

and

$$\text{div} \, \mathbf{E} = \frac{\rho}{\varepsilon_0}. \tag{9.54}$$

Change all the differential operators from Σ to Σ' according to Table 9.2 in equations (9.53) and (9.54), use $c^2 = 1/\varepsilon_0 \mu_0$ and $\gamma^2 (1 - \beta^2) = 1$. Eliminating $(\partial E_z / \partial z')$ between them gives

$$\frac{\partial}{\partial x'} [\gamma (c B_y - \beta E_x)] - \frac{\partial}{\partial y'} [\gamma (c B_x - \beta E_y)] = \mu_0 c \gamma (j_z - \beta c \rho) + \frac{\partial E_z}{\partial ct'}.$$

This is

$$(\text{curl}' \, c\mathbf{B}')_{z'} = \mu_0 c j'_{z'} + \frac{\partial E'_{z'}}{\partial ct'}$$

if, in addition to the fields $\mathbf{E}$ and $\mathbf{B}$ satisfying the transformation equations (9.44) and (9.52), we also have

$$j'_{z'} = \gamma (j_z - \beta c \rho). \tag{9.55}$$

Similarly eliminating $\partial E_z / \partial ct'$ instead of $\partial E_z / \partial ct$ gives an equation which is the same as

$$\text{div}' \, \mathbf{E}' = \frac{\rho'}{\varepsilon_0}$$

if

$$\rho' c = \gamma (\rho c - \beta j_z). \tag{9.56}$$

Transforming in a similar manner the equations for the x- and y-components of curl $\mathbf{B}$ shows that

$$j'_{x'} = j_x \quad \text{and} \quad j'_{y'} = j_y. \tag{9.57}$$

Equations (9.55)–(9.57) are the same as would be the case if $(\rho c, \mathbf{j})$ transformed as does $(ct, \mathbf{r})$. Therefore $(\rho c, \mathbf{j})$ is a four-vector.

We have confirmed the transformation equations (9.44) for the **E** and **B** fields by specifically requiring the covariance of Maxwell's equations. In addition, we have found that the charge density and the three components of the current density may be appropriately combined into a four-vector.

+9.10　The electromagnetic four-vector potential

The three-vector potential **A** has the property that

$$\mathbf{B} = \text{curl } \mathbf{A}. \tag{9.58}$$

In addition the electric field is then given by

$$\mathbf{E} = -\text{grad } \phi - \frac{\partial \mathbf{A}}{\partial t}, \tag{9.59}$$

where ϕ is the scalar electrostatic potential. We can take a step towards discovering the special relativity in these two equations by making the units manifestly uniform. Since **E** and $c\mathbf{B}$ have the same dimensions it follows that ϕ and $c\mathbf{A}$ also have the same dimensions so, for example, equation (9.59) becomes

$$\mathbf{E} = -\text{grad } \phi - \frac{\partial c\mathbf{A}}{\partial ct}.$$

Suppose this equation applies in frame Σ. Take the z component of this equation and change all the differential operators from Σ to their equivalents in Σ' (Table 9.2). The result is

$$E_z = -\frac{\partial}{\partial z'}[\gamma(\phi - \beta c A_z)] - \frac{\partial}{\partial ct'}[\gamma(c A_z - \beta\phi)].$$

But $E'_{z'} = E_z$ and we have covariance if

$$\phi' = \gamma(\phi - \beta c A_z) \quad \text{and} \quad c A'_{z'} = \gamma(c A_z - \beta\phi).$$

These two equations show that ϕ and $c A_z$ transform as do ct and z. By transforming components of equation (9.58) and using the transformation properties of **E** and **B** it is possible to show that

$$A'_{x'} = A_x \quad \text{and} \quad A'_{y'} = A_y.$$

Thus $(\phi, c\mathbf{A})$ transforms as does $(ct, \mathbf{r})$ and is therefore a four-vector.

There is a comment to be made about the four-vector potential. The observed fields are **E** and **B**. Given equations (9.58) and (9.59) it is easy to show that these fields are unchanged if $(\phi, c\mathbf{A})$ is changed as follows:

$$(\phi, c\mathbf{A}) \rightarrow (\phi - \frac{\partial \Lambda}{\partial ct}, c\mathbf{A} + \text{grad}\Lambda),$$

where Λ is a scalar function of $(ct, \mathbf{r})$. This kind of change is called a **gauge transformation** and the fact that it make no difference to the observable fields is called **gauge invariance**. This important property is beyond the scope of our subject, except in one respect in the next paragraph.

Since the vector differentials of the **E** and **B** fields at a point depend on sources of charge and current at that point, the same must be true for the four-vector $(\phi, c\mathbf{A})$. So we have to relate this four-vector to the four-vector $(\rho c, \mathbf{j})$. We know from simple electrostatics that the electrostatics potential ϕ can be non-zero at points where there is no charge ($\rho = 0$). In fact, in electrostatics we have

$$\text{div grad } \phi = -\frac{\rho}{\varepsilon_o}. \tag{9.60}$$

Now div grad($\equiv \nabla^2$) is not a relativistically invariant operator but $\nabla^2 - (\partial^2/\partial(ct)^2)$ is (equation (9.76)). This suggests that it is the equation

$$\left(\nabla^2 - \frac{1}{c^2} \frac{\partial^2}{\partial t^2} \right) \phi = -\frac{\rho}{\varepsilon_o}$$

that is correct. It connects the first component of one four-vector to the first component of another. Why should it not connect the remaining components in the same way? So we write

$$\left(\nabla^2 - \frac{1}{c^2} \frac{\partial}{\partial t^2} \right) (\phi, c\mathbf{A}) = -\sqrt{\frac{\mu_o}{\varepsilon_o}} (\rho c, \mathbf{j}). \tag{9.61}$$

(Keep in mind that $c = 1/\sqrt{\varepsilon_0 \mu_0}$.) This last equation is actually four equations connecting two four-vectors component-by-component. Since the operator on the left is invariant, the equation is clearly Lorentz covariant. (Sorry about all the c's and other constants. This is a result of keeping the dimensions of all the components of each four-vector the same yet retaining the conventional symbols for physical quantities, and not putting $c = 1$! And the equation is correct.)

There is one problem for equation (9.61): it is only true in gauges that make

$$\text{div } c\mathbf{A} + \frac{\partial \phi}{\partial ct} = 0. \tag{9.62}$$

This is also a Lorentz covariant equation so that the validity of equation (9.61) is not affected by a Lorentz transformation. See Jackson[9] for a discussion of this matter.

Equation (9.61) shows that in a vacuum the four-vector potential satisfies a wave equation with a propagation speed equal to that of light. This is no surprise. It also means that the solutions of these equations for fields at one point when the sources are elsewhere have to take account of the time it takes source changes to propagate their effect to such a distant point.

Use a result from this section to attempt Problem 9.14.

---

9 Jackson, *op. cit.*, Section 6.5, pp. 220–3.

$^{+}$9.11   Tensorland

Since introducing four-vectors we have been moving towards the use of a technique and notation that fits physical quantities into mathematical objects that make the properties of these quantities under Lorentz transformations clear. An example is tailoring energy and momentum into a four-vector. Once these objects have been identified it is easy to construct equations in one inertial reference frame secure in the knowledge that they will be true in all inertial frames. The technique is that of **tensors**.

For a straight forward introduction to tensors our readers are referred to the book by Boas.[10] Here we shall cover sufficient material we trust, to convince, the reader of the usefulness of the techniques of tensors and to hint at their importance in general relativity. However this approach is, by modern standards, not as 'geometrical' as is now favoured.

Let us dispose of some vocabulary and notation. A tensor T will, in general, have many components and at a point in space–time each component is specified by the integer values of some **indices**. Thus T^{12}_{302} is a particular component of the tensor T. However, we do not normally have to pick out such a particular component and the general component might be written $T^{\delta\varepsilon}_{\alpha\beta\gamma}$. As is the case for a vector, a tensor will, in general, have different components in different inertial frames.

1   In relativity we have a four-dimensional space–time, so each index may have one of four possible values, 0, 1, 2, or 3. For example, x^{μ}, $\mu = 0, 1, 2, 3$ for the four-vector (= a rank-1 contravariant tensor) that is the time–space coordinates of an event. We shall stay with **dimension** $n = 4$, lower case Greek letters to represent the indices, and imply the above integer range for the indices.

2   The number of indices carried by a tensor is called the **rank**. It is 5 for T above. Therefore in an n-dimensional space a rank-r tensor has n^r components. That is, $4^5 = 1024$ for T.

3   If all indices are in the superscript position the tensor is said to be a **contravariant** tensor. For example the tensor S with components $S^{\delta\varepsilon}$.

4   If all indices are in the subscript position the tensor is said to be a **covariant** tensor. For example, the tensor R with components $R_{\alpha\beta\gamma}$.

5   If some of the indices are up and some are down the tensor is said to be **mixed**. For example, the tensor T with components $T^{\delta\varepsilon}_{\alpha\beta\gamma}$. A rank zero tensor has no indices and therefore has only one component. A *scalar* is another name for such a tensor.

6   In the text we shall normally name a tensor (T) with its indices attached ($T^{\delta\varepsilon}_{\alpha\beta\gamma}$) without thereby implying a single component.

There is an operational shorthand that is very useful. It is called the **Einstein summation convention**. If the same symbol occurs once in the contravariant index

---

10 Boas, M. L., *Mathematical methods in the physical sciences*, 2nd edn, New York: John Wiley & Sons, Chapter 10.

position and once in the covariant index position in one term of an expression, this means summing over the n possible values of that symbol. Thus for $n = 4$ we have

$$T^{\alpha\varepsilon}_{\alpha\beta\gamma} = \sum_{\alpha=0}^{3} T^{\alpha\varepsilon}_{\alpha\beta\gamma}.$$

Such an index is called a **dummy index** as the actual symbol has no significance. Such a summation performed on a single tensor, which must be mixed, is called **contraction** and produces a new tensor of rank less by two.

Tensors are properly defined by their properties under transformations. In special relativity that means the Lorentz transformations. We remind readers that under the change from Σ to Σ' (Section 7.2) we have that the interval $\mathrm{d}ct$, $\mathrm{d}x$, $\mathrm{d}y$, $\mathrm{d}z$ between an event and another infinitesimally close transforms as a four-vector:

$$\begin{pmatrix} \mathrm{d}ct' \\ \mathrm{d}x' \\ \mathrm{d}y' \\ \mathrm{d}z' \end{pmatrix} = \begin{pmatrix} \gamma & 0 & 0 & -\gamma\beta \\ 0 & 1 & 0 & 0 \\ 0 & 0 & 1 & 0 \\ -\gamma\beta & 0 & 0 & \gamma \end{pmatrix} \begin{pmatrix} \mathrm{d}ct \\ \mathrm{d}x \\ \mathrm{d}y \\ \mathrm{d}z \end{pmatrix}.$$

We can generalise and extend this in a tidy manner.

1   Use our four-vector notation so that, for example, $(\mathrm{d}ct, \mathrm{d}\mathbf{r})$ appears as $\mathrm{d}x^0$, $\mathrm{d}x^1$, $\mathrm{d}x^2$, $\mathrm{d}x^3$.

2   Use the chain rule for the infinitesimals, for example,

$$\mathrm{d}x'^\alpha = \frac{\partial x'^\alpha}{\partial x^0}\mathrm{d}x^0 + \frac{\partial x'^\alpha}{\partial x^1}\mathrm{d}x^1 + \frac{\partial x'^\alpha}{\partial x^2}\mathrm{d}x^2 + \frac{\partial x'^\alpha}{\partial x^3}\mathrm{d}x^3.$$

3   Simplify the notation with the Einstein summation convention, so that, for example,

$$\mathrm{d}x'^\alpha = \frac{\partial x'^\alpha}{\partial x^\mu}\mathrm{d}x^\mu, \tag{9.63}$$

which, for convenience, we write

$$\mathrm{d}x'^\alpha = \Lambda^\alpha_\mu\, \mathrm{d}x^\mu.$$

Note that we are using the Einstein convention although Λ is not a tensor. (It bridges two inertial frames and its elements cannot be constructed in a single inertial frame.) Equation (9.63) is correct for any linear transformation of the interval. To make it into a Lorentz transformation the partial differentials have to satisfy certain relations to ensure compliance with Einstein's postulates (Section 2.4). See Weinberg.[11]

11 Weinberg, S., *Gravitation and cosmology: Principles and applications of the general theory of relativity*, 1972, New York: John Wiley & Sons, Section 2.1.

4   The reverse transformation is clearly

$$dx^\mu = \frac{\partial x^\mu}{\partial x'^\beta} dx'^\beta = \bar\Lambda^\mu_\beta \, dx'^\beta. \tag{9.64}$$

5   The there-and-back transformation is

$$dx^\mu = \bar\Lambda^\mu_\beta \Lambda^\beta_\gamma \, dx^\gamma.$$

This only makes sense if

$$dx^\mu = \delta^\mu_\gamma \, dx^\gamma$$

where $\delta^\mu_\gamma = 1$ if $\mu = \gamma$ and zero otherwise. Therefore

$$\bar\Lambda^\mu_\beta \Lambda^\beta_\gamma = \frac{\partial x^\mu}{\partial x'^\beta} \frac{\partial x'^\beta}{\partial x^\gamma} = \delta^\mu_\gamma. \tag{9.65}$$

The last equality in equation (9.65) is true in general for the partial differentials involved. *If* we want to think of $\bar\Lambda$ and Λ as matrices, this equation means that each is the inverse of the other.

Now we can define the transformation rule for the components of a tensor. (We use the symbol T with indices for the mixed tensor in a given inertial frame that we use as an example. In another inertial frame it is still the same tensor but its components are changed and for that we shall use T' with indices.)

1   Under a transformation in which a four-vector of the interval transforms as in equation (9.63) and transforms back again as in equation (9.64) we have that . . .
2   Each contravariant index (for example δ) on T becomes a contravariant index (v) on T' by a Λ^v_δ factor and an Einstein summation, as in the transformation of the components of a vector, equation (9.63) and . . .
3   Each covariant index (α) on T becomes a covariant index (κ) on T' by a $\bar\Lambda^\alpha_\kappa$ factor and a summation, as in the reverse transformation of the components of a vector, equation (9.64).

Therefore, for our example,

$$T \to T', \quad T'^{v\rho}_{\kappa\lambda\mu} = \bar\Lambda^\alpha_\kappa \bar\Lambda^\beta_\lambda \bar\Lambda^\gamma_\mu \Lambda^v_\delta \Lambda^\rho_\varepsilon T^{\delta\varepsilon}_{\alpha\beta\gamma}. \tag{9.66}$$

Notice five summations over repeated indices on the right-hand side! And, of course, we see that a four-vector is a contravariant, rank-1 tensor.

Now some rules about operations with tensors:

1   An equality between two tensors means a component-by-component equality and therefore must be an equality between tensors of identical dimension, rank and mix. (We call *dimension, rank* and *mix* the tensor *type*.)

2   Summing and differencing of tensors may only be performed with tensors of the same type.

3   There can be null tensors. These have all components zero.

4   The **direct product** of two tensors involves multiplying each component of one by the entire assembly of the other tensor. This is possible only if the tensors have the same dimension n and it produces a tensor of rank equal to the sum of the two ranks: $R_{\alpha\beta\gamma} S^{\delta\varepsilon} = T^{\delta\varepsilon}_{\alpha\beta\gamma}$.

5   Contraction may also be performed on the direct product of two tensors, thus:

$$R^{\delta\varepsilon}_{\alpha\beta\gamma} S^{\alpha\rho} = \sum_{\alpha=0}^{3} R^{\delta\varepsilon}_{\alpha\beta\gamma} S^{\alpha\rho} = P^{\delta\varepsilon\rho}_{\beta\gamma},$$

where P is also a tensor. The reader will be able to prove that contraction produces a tensor by contracting both sides of equation (9.66) and using equation (9.65). (See Problem 9.17.)

The **metric tensor** $g_{\mu\nu}$ is important and we introduce it by considering the square of the interval (see Section 7.6). In generalised coordinates it is defined by:

$$(ds)^2 = g_{\mu\nu}dx^\mu dx^\nu. \tag{9.67}$$

However, in Cartesian coordinates as we have used them for our inertial frames we have

$$(ds)^2 = (dct)^2 - (dx)^2 - (dy)^2 - (dz)^2$$
$$= (dx^0)^2 - (dx^1)^2 - (dx^2)^2 - (dx^3)^2.$$

Therefore

$$\begin{aligned}
g_{\mu\nu} &= 1 & \text{if } \mu = \nu = 0, \\
g_{\mu\nu} &= -1 & \text{if } \mu = \nu = 1, 2 \text{ or } 3, \\
g_{\mu\nu} &= 0 & \text{if } \mu \neq \nu.
\end{aligned} \tag{9.68}$$

We know $(ds)^2$ is an invariant ($=$ a rank-0 tensor) so in equation (9.67) $g_{\mu\nu}$ connects a rank-0 tensor to the direct product of two rank-1 contravariant tensors. Therefore $g_{\mu\nu}$ is itself a rank-2 covariant tensor. There is also a contravariant partner to $g_{\mu\nu}$, which we denote by $g^{\mu\nu}$. It has the same components as does $g_{\mu\nu}$ for the same indices. But it is not identical to $g_{\mu\nu}$ because it is contravariant and the latter is covariant! Many books use a Greek η instead of g for the metric tensor in the inertial reference frames we have used. This is to stress that the metric in flat space–time is like a simple matrix of zeros and $\pm$ ones whereas g in curved space can be much less simple.

Thus the properties of $g_{\mu\nu}$ and $g^{\mu\nu}$ summarised are:

i $g_{\mu\nu}$ and $g^{\mu\nu}$ are tensors.

ii   The components of $g_{\mu\nu}$ are the same in all inertial frames when the space coordinate system is Cartesian. The same is true for $g^{\mu\nu}$. (See Problem 9.15.)

iii $g_{\mu\nu}$ serves to change a tensor into another tensor with one less contravariant index and one more covariant index, thus:

$$g_{\mu\nu}T^{\nu\varepsilon}_{\alpha\beta\gamma} = R^{\varepsilon}_{\mu\alpha\beta\gamma}.$$

This is called **lowering an index**. Since $g_{\mu\nu}$ and T are tensors so is R.

iv   Similarly $g^{\mu\nu}$ abolishes a covariant index and creates a contravariant index to produce a new tensor:

$$g^{\mu\nu}T^{\delta\varepsilon}_{\nu\beta\gamma} = S^{\mu\delta\varepsilon}_{\beta\gamma}.$$

This is called **raising an index**.

v   Since

$$(\mathrm{d}s)^2 = g_{\mu\nu}\,\mathrm{d}x^\mu\,\mathrm{d}x^\nu = \mathrm{d}x_\nu\,\mathrm{d}x^\nu \tag{9.69}$$

the object $\mathrm{d}x_\nu$ is a component of a covariant version of the contravariant vector of which $\mathrm{d}x^\nu$ is a component. Now

$$\mathrm{d}x_\nu = g_{\nu\mu}\mathrm{d}x^\mu$$

then using equations (9.68)

$$(\mathrm{d}x_0, \mathrm{d}x_1, \mathrm{d}x_2, \mathrm{d}x_3) = (g_{00}\mathrm{d}x^0, g_{11}\mathrm{d}x^1, g_{22}\mathrm{d}x^2, g_{33}\mathrm{d}x^3)$$
$$= (\mathrm{d}x^0, -\mathrm{d}x^1, -\mathrm{d}x^2, -\mathrm{d}x^3)$$
$$= (\mathrm{d}ct, -\mathrm{d}\mathbf{r}). \tag{9.70}$$

Thus to every contravariant vector there is a covariant vector, and the other way round. For example, $(ct, \mathbf{r})$ and $(ct, -\mathbf{r})$, $(E, \mathbf{P}c)$ and $(E, -\mathbf{P}c)$.

vi   The *inner product* (equation (8.37)) of a contravariant vector and a covariant vector is an invariant. Thus:

$$x \cdot y = x^\mu y_\mu = x^0 y_0 + x^1 y_1 + x^2 y_2 + x^3 y_3$$
$$= g_{\mu\nu}\,x^\mu y^\nu = x^0 y^0 - x^1 y^1 - x^2 y^2 - x^3 y^3$$
$$= g^{\mu\nu}\,x_\mu y_\nu = x_0 y_0 - x_1 y_1 - x_2 y_2 - x_3 y_3. \tag{9.71}$$

This may appear to be trivial. The metric tensor is this very simple object in our flat space–time with its time and Cartesian coordinates. In General Relativity the role of the metric tensor is to define the invariant interval, as in equation (9.69), at a point in curved, non-Euclidean space–time. It then carries information about the curvature of space–time at that point.

Let us now look at the differential operators $\partial/\partial ct$, $\partial/\partial x$, $\partial/\partial y$, $\partial/\partial z = \partial/\partial x^\mu$, $\mu = 0, 1, 2, 3$. These transform as

$$\frac{\partial}{\partial x'^\beta} = \frac{\partial x^\mu}{\partial x'^\beta}\frac{\partial}{\partial x^\mu}. \tag{9.72}$$

Compare this with the transformation of a covariant four-vector U_μ that will be, according to equation (9.66):

$$U'_\beta = \frac{\partial x^\mu}{\partial x'^\beta}U_\mu. \tag{9.73}$$

Thus U_μ and $\partial/\partial x^\mu$ transform in the same way and therefore the latter is also a covariant vector. For this reason the following shortcut notation is used:

$$\frac{\partial}{\partial x^\mu} = \partial_\mu. \tag{9.74}$$

The position of the index is fixed as required for a covariant vector. Similarly the differentials with respect to the covariant coordinates are contravariant and the following notation is used:

$$\frac{\partial}{\partial x_\nu} = \partial^\nu. \tag{9.75}$$

At this point the reader should spare a few moments to read Table 9.3 concerning the differentiation of tensors. The difficulty mentioned does not apply in our usual arrangement of inertial frames. With this caveat the properties of our differential operators follow:

1   $\partial_\mu = g_{\mu\nu}\partial^\nu$ and $\partial^\mu = g^{\mu\nu}\partial_\nu$.
2   The quantities such as $\partial_\mu\phi$ (ϕ is a scalar = a rank-0 tensor), $\partial^\nu\phi$, $\partial^\mu\partial_\nu\phi$, are tensors of type determined by the number and disposition of the indices.
3   The quantity $g_{\mu\nu}\partial^\mu\partial^\nu$ must be an invariant. In fact

$$-g_{\mu\nu}\partial^\mu\partial^\nu = -\partial^\mu\partial_\mu = \frac{\partial^2}{\partial x^2} + \frac{\partial^2}{\partial y^2} + \frac{\partial^2}{\partial z^2} - \frac{1}{c^2}\frac{\partial^2}{\partial t^2} \tag{9.76}$$

$$= \square \quad \text{(the d'Alembertian operator)}.$$

4   It follows that quantites such as $\square\phi$, $\square A^\mu$ are also tensors typed by their indices or by their lack of indices.

Those properties allow us to reformulate the equations of electromagnetism in a fully covariant manner. We shall do that in Section 9.12.

It is worth comparing features of vector analysis in three dimensions, a subject that must now be familiar to most readers, to tensor analysis in more dimensions. Students are introduced to a vector through statements such as

$$\mathbf{V} = V_x\mathbf{i} + V_y\mathbf{j} + V_z\mathbf{k},$$

Table 9.3 The differentiation of tensors

We have used the idea of a vector in a variety of circumstances. Examples are the vector connecting two points or the momentum of a particle. However, there are also vector fields for which the vector has a value at all points within some finite or infinite space. Examples are the electric field or the velocity field of a moving fluid. We can now add tensor fields. Given such a field, the tensor may vary with position and time and the derivatives of the field with respect to these variables may be of importance, as in Maxwell's equations. In this table we examine the differentiation of a contravariant vector field as an example of the difficulty that can occur. Consider equation (9.63) for the transformation of the components of contravariant vector:

$$V'^{\alpha} = \frac{\partial x'^{\alpha}}{\partial x^{\mu}} V^{\mu}.$$

Partially differentiate both sides with respect to the coordinate x'^{β} putting

$$\frac{\partial}{\partial x'^{\beta}} = \frac{\partial x^{\nu}}{\partial x'^{\beta}} \frac{\partial}{\partial x^{\nu}}$$

on the right-hand side:

$$\frac{\partial V'^{\alpha}}{\partial x'^{\beta}} = \frac{\partial}{\partial x'^{\beta}} \left[\frac{\partial x'^{\alpha}}{\partial x^{\mu}} V^{\mu} \right] = \frac{\partial x^{\nu}}{\partial x'^{\beta}} \frac{\partial}{\partial x^{\nu}} \left[\frac{\partial x'^{\alpha}}{\partial x^{\mu}} V^{\mu} \right].$$

We can immediately see a problem. One of the two terms that appears from the differentiation contains a factor

$$\frac{\partial^2 x'^{\alpha}}{\partial x^{\nu} \partial x^{\mu}}.$$

The transformations between our usual arrangement of inertial frames in special relativity are linear so that this second-order derivative is zero, What remains is

$$\frac{\partial V'^{\alpha}}{\partial x'^{\beta}} = \frac{\partial x^{\nu}}{\partial x'^{\beta}} \frac{\partial x'^{\alpha}}{\partial x^{\mu}} \frac{\partial V^{\mu}}{\partial x^{\nu}} = \bar{\Lambda}^{\nu}_{\beta} \Lambda^{\alpha}_{\mu} \partial_{\nu} V^{\mu}.$$

Comparing this with equations (9.63) and (9.66) we see that this is the equation for the transformation of a rank-2 mixed tensor, $\partial_{\nu} V^{\mu}$ in this case. This is a desirable feature for the derivative of a vector or, in a more general way, for any tensor. All this is correct in Cartesian-based reference frames but is not true if that second-order differential is not zero.

When does this occur? The answer is if the elements of Λ and $\bar{\Lambda}$ vary with position. In three-dimensional space that happens with curvilinear coordinate systems and in four dimensions with curved space–time. But all that is outside the scope of this book.

---

where **i**, **j**, and **k** are unit vectors along the x, y, and z axes of Cartesian coordinates. These unit vectors are called **basis vectors**. Other sets of basis vectors are possible, for example, three orthogonal unit vectors at a point along directions defined by increasing r, θ, and ϕ in spherical polar coordinates. We have not discussed that aspect of tensor analysis although it will be clear that we have given the transformations in the context of a change of basis from unit vectors in one inertial frame to those in another inertial frame.

Vector calculus in three dimensions takes the student quickly to a situation in which vector relations are written in index-free notation. Thus Maxwell's equations are correct in all inertial frames. In the next section we shall rewrite these equations using tensor relations. Although they are then manifestly relativistically covariant they will not be index-free. A reader might well ask if tensor notation can be made index-free. The answer is yes, but it requires considerable mathematical development that is not appropriate to this book. The subject becomes *differential geometry*. The geometry aspect is perhaps illustrated by the interpretation of contravariant and covariant vectors. The latter are called 1-forms and are functions for which a (contravariant) vector is the argument. Geometrically the 1-form may be visualised as a set of planes in space a certain number of which are pierced by the vector. The result of a particular 1-form having a particular vector as its argument is a real number equal to the number of planes pierced by the vector. In tensor language that number is the inner product of the contravariant and the covariant vector.

Books by Schutz[12] and by Misner, Thorne and Wheeler[13] employ this approach in the context of relativity. For a more general introduction to differential geometry see another book by Schutz.[14]

For practice with tensors try Problems 9.15–9.17.

+9.12   Electromagnetism with tensors

Let us start from the four-vector potential $(\phi, c\mathbf{A})$ that we met in Section 9.10. Here ϕ is the scalar potential that we know as the electrostatic potential in time-independent electromagnetism, and $\mathbf{A}$ is the vector potential. Since we have to work with tensors we represent this four-vector by the contravariant A^μ. The covariant vector is $A_\mu = g_{\mu\nu} A^\nu = (\phi, -c\mathbf{A})$. We shall sometimes name such vectors without their indices. The equations (9.58) and (9.59).

$$\mathbf{B} = \operatorname{curl} \mathbf{A} \quad \text{and} \quad \mathbf{E} = -\operatorname{grad} \phi - \frac{\partial \mathbf{A}}{\partial t}$$

may be put together to give six terms such as

$$cB_x = c\left(\frac{\partial A_z}{\partial y} - \frac{\partial A_y}{\partial z}\right) = -\partial_2 A_3 + \partial_3 A_2,$$

$$E_x = -\frac{\partial \phi}{\partial x} - \frac{\partial A_x}{\partial t} = -\partial_1 A_0 + \partial_0 A_1. \tag{9.77}$$

12 Schutz, B. F., *A first course in general relativity*, Cambridge: Cambridge University Press, 1985 (reprinted 1992).
13 Misner, C. W., Thorne, K. S., and Wheeler, J. A., *Gravitation*, San Francisco: W. H. Freeman & Co., 1970.
14 Schutz, B. F., *Geometrical methods of mathematical physics*, Cambridge: Cambridge University Press, 1980.

These equations express some changes from $(ct, \mathbf{r})$ and $(\phi, c\mathbf{A})$ notation to tensor notation, x^μ and A_ν. So, the six field quantities $\mathbf{E}$ and $\mathbf{B}$ are given by the four-dimensional curl of the four-vector A. If we define F by

$$F_{\mu\nu} = \partial_\mu A_\nu - \partial_\nu A_\mu \tag{9.78}$$

then, by the rules applying to tensors, the right-hand side and hence the left-hand side are rank-2 covariant tensors. F is sometimes called the **Faraday**. The components in Cartesian coordinates are given by the array

$$
\begin{array}{c}
\nu = \quad 0 \qquad 1 \qquad 2 \qquad 3 \\
\\
\mu = \\
F_{\mu\nu} =
\begin{array}{cccc}
0 & E_x & E_y & E_z \\
-E_x & 0 & -cB_z & cB_y \\
-E_y & cB_z & 0 & -cB_x \\
-E_z & -cB_y & cB_x & 0
\end{array}
\quad
\begin{array}{c}
0 \\
1 \\
2 \\
3
\end{array}
\end{array} \tag{9.79}
$$

The properties of F are as follows:

1   The transformation of $F_{\mu\nu}$ from inertial frame Σ to $F'_{\mu\nu}$ in Σ' of our standard arrangement is given by applying the rule of equation (9.66) with appropriate coefficients in the Λ matrices. This gives the same results as found in Section 9.8 and expressed in equations (9.44). Problem 9.18 asks the reader to check these statements.

2   A more general Lorentz transformation may be constructed, if required, by the use of equation (9.63) to obtain the elements of the Λ needed.

3   The related tensors F^μ_ν and $F^{\mu\nu}$ may be constructed from $F_{\mu\nu}$ by using the index raising ability of $g^{\mu\nu}$.

4   Note that $F_{\mu\nu}$ and $F^{\mu\nu}$ are anti-symmetric, that is $F_{\nu\mu} = -F_{\mu\nu}$ and $F^{\nu\mu} = -F^{\mu\nu}$.

We have to connect the tensor F to the sources of the fields. Here we need the four-vector of current density $(\rho c, \mathbf{j})$ where ρ is the electric charge volume density, and $\mathbf{j}$ is the electric current density (Section 9.9). We represent this vector by the contravariant vector $J^\nu = (\rho c, \mathbf{j})$. The covariant form is $J_\nu = g_{\nu\mu} J^\mu = (\rho c, -\mathbf{j})$. We can make the required connection between rank-2 F and rank-1 J by differentiating F and contracting a pair of indices, thus:

a $\quad \partial^\mu F_{\mu\nu} = \sqrt{\dfrac{\mu_0}{\varepsilon_0}} J_\nu \tag{9.80}$

gives the equations

$$\text{curl } \mathbf{B} = \mu_0 \left(\mathbf{j} + \varepsilon_0 \frac{\partial \mathbf{E}}{\partial t} \right) \quad \text{and} \quad \text{div } \mathbf{E} = \rho/\varepsilon_0.$$

Also:

b   From equation (9.78) it is easy to show that

$$\partial_\kappa F_{\mu\nu} + \partial_\mu F_{\nu\kappa} + \partial_\nu F_{\kappa\mu} = 0, \tag{9.81}$$

where κ, λ, and μ are all different. The choice of $\kappa = 1$, $\lambda = 2$, and $\mu = 3$ gives the equation div $\mathbf{B} = 0$, as required by the absence of magnetic monopoles. But the choice of three indices from four means that there are four equations in equation (9.81). The remaining three, each with one of the indices set equal to zero, must also be satisfied. They give the three equations

$$\text{curl } \mathbf{E} = -\frac{\partial \mathbf{B}}{\partial t}.$$

c   Since $\partial^\mu \partial^\nu = \partial^\nu \partial^\mu$ and $F_{\mu\nu}$ is anti-symmetric we have $\partial^\nu \partial^\mu F_{\mu\nu} = 0$. Therefore

$$\partial^\nu \partial^\mu F_{\mu\nu} = \sqrt{\frac{\mu_0}{\varepsilon_0}} \partial^\nu J_\nu = 0. \tag{9.82}$$

The last equality is the continuity equation (9.36), div $\mathbf{j} + \partial\rho/\partial t = 0$. We leave the reader to confirm these results by completing Problem 9.19.

The next operation is to find the differential equations satisfied by the components of the four-vector potential A. We have equation (9.78)

$$F_{\mu\nu} = \partial_\mu A_\nu - \partial_\nu A_\mu$$

and (9.80)

$$\partial^\mu F_{\mu\nu} = \sqrt{\frac{\mu_0}{\varepsilon_0}} J_\nu.$$

Therefore

$$\sqrt{\frac{\mu_0}{\varepsilon_0}} J_\nu = \partial^\mu \partial_\mu A_\nu - \partial^\mu \partial_\nu A_\mu.$$

But $\partial^\mu \partial_\nu = \partial_\nu \partial^\mu$ and in any Lorentz gauge $\partial^\mu A_\mu = 0$. (This is equation (9.62) written in tensor notation.) Therefore in such a gauge we have (using the definition of the d'Alembertian operator $\square$, equation (9.76))

$$-\partial^\mu \partial_\mu A_\nu = \square A_\nu = -\sqrt{\frac{\mu_0}{\varepsilon_0}} j_\nu. \tag{9.83}$$

This is the same as the four equations of (9.61) that we proposed for the four components of $(\phi, c\mathbf{A})$:

$$\left(\nabla^2 - \frac{1}{c^2}\frac{\partial^2}{\partial t^2}\right)(\phi, c\mathbf{A}) = -\sqrt{\frac{\mu_0}{\varepsilon_0}}(\rho c, \mathbf{j}).$$

The reader might well complain that proposing a vector field A and then juggling with tensor operations is not necessarily in contact with the reality of the experimental experience of electricity and magnetism. But as soon as we connected F

to J in equation (9.80) we introduced electric charge and current, and allowed relativistic covariance to dictate the relation between all these quantities. The factor that relativity did not determine was the constant in front of J in equation (9.80). To be consistent with the quantitative definition of the electric and magnetic fields that constant had to be $\mu_0 c = \sqrt{\mu_0/\varepsilon_0}$. In addition, equation (9.81) allowed for the observed absence of magnetic monopoles. The total result is that equations (9.80) and (9.81) are a compact statement of all Maxwell's equations (in a vacuum with sources) and equation (9.83) is equivalent to these in that it gives the differential equations satisfied by the potentials ϕ and $\mathbf{A}$. The experimental results are there to be found. For example, in static conditions, the $\nu = 0$ component of equation (9.83) yields the equation

$$\Box A_0 = \nabla^2 \phi = -\rho/\varepsilon_0$$

This is Poisson's equation, a well-known result for the electrostatic potential in terms of the charge density.

In addition, if $\bar{v}$ is the four-velocity $(\gamma_v c, \gamma_v \mathbf{v})$ of a particle carrying charge q and p is its four-momentum then the equation

$$\frac{dp^\lambda}{d\tau} = q g^{\lambda\mu} F_{\mu\nu} \bar{v}^\nu = q F^\lambda_\nu \bar{v}^\nu \tag{9.84}$$

is the relativistically covariant equation for the proper time rate of change for the four-momentum of a charged particle moving in electric and magnetic fields. We constructed that equation in Section 9.8 from equation (5.20). The right-hand side of equation (9.84) is a contravariant four-vector of force, $\bar{f}$ (a Minkowski force, as in equation (8.42)).

Writing Maxwell's equations for the electromagnetic field in the vacuum with sources in this relativistically covariant form illustrates two points. The first is that given the possibility of a four-vector field (A^μ), coupled solely to electric charge, relativity constrains the theoretical description of the field to that physically realised by nature. The second point is that tensors provide a tool well-suited to the development of relativity. And, as we have hinted, that usefulness extends to general relativity so that this introduction to tensors may serve the reader in undertaking a study of that subject. But, beware, there are notational variations.

Summarising, Maxwell's equations may be written (including a tensorial shorthand from Problem 9.20):

$$\partial^\mu F_{\mu\nu} = \sqrt{\frac{\mu_0}{\varepsilon_0}} J_\nu,$$

$$\varepsilon^{\kappa\lambda\mu\nu} \partial_\lambda F_{\mu\nu} = 0.$$

The differential equation satisfied by the vector potential is

$$-\Box A_\nu = \sqrt{\frac{\mu_0}{\varepsilon_0}} J_\nu.$$

The effect of the Lorentz force is contained in

$$\frac{\mathrm{d}p^{\lambda}}{\mathrm{d}\tau} = q F^{\lambda}_{v} \bar{v}^{v}.$$

*9.13   What to remember

The reader who has reached this point will probably know the equations upon which is based all the material we have presented. As a reminder we give here the equations that are worth remembering. Five of them (not in vector notation) were already marked with an asterisk in Table 2.2. We quote them again in vector notation and add one more.

1   Equation (2.3) $\mathbf{P} = M\mathbf{v}/\sqrt{1 - v^2/c^2}$.
2   Equation (2.5) $E = Mc^2/\sqrt{1 - v^2/c^2}$.
3   Equation (3.1) $M^2 c^4 = E^2 - \mathbf{P} \cdot \mathbf{P} c^2$ (for any massive particle or a set of particles as a whole).
4   Equation (3.2) $\mathbf{v} = \mathbf{P}c^2/E$.
5   Equation (2.8) $E = \mathbf{P}c$, for a photon.
6   Equation (8.20)   To transform a four-vector $\overline{V}'$ in Σ' to $\overline{V}$ in Σ:

$$\overline{V} = \mathbf{L}(\beta_u)\overline{V}'$$

where $\overline{V}$ and $\overline{V}'$ are expressed as column matrices,

$$\mathbf{L}(\beta_u) = \begin{bmatrix} \gamma_u & 0 & 0 & \gamma_u \beta_u \\ 0 & 1 & 0 & 0 \\ 0 & 0 & 1 & 0 \\ \gamma_u \beta_u & 0 & 0 & \gamma_u \end{bmatrix}$$

and $\beta_u = u/c$, $\gamma_u = 1/\sqrt{1 - \beta_u^2}$.

We also list the important four-vectors that we have used:
a   Coordinates of an event. $x = (ct, \mathbf{r})$ (Section 8.10).
b   Four-velocity of a particle or $\bar{v} = (\gamma_v c, \gamma_v \mathbf{v})$ (equation (8.39)).
    wave group
c   Energy–momentum $p = (E, \mathbf{P}c)$ (equation (8.40)).
d   Four-acceleration $\bar{a} = \mathrm{d}\bar{v}/\mathrm{d}\tau$ (equation (8.41)).
e   Wave four-vector $k = (\omega/c, \mathbf{k})$ (equation(8.43)).
f   Charge and current density $J = (\rho c, \mathbf{j})$ (Section 9.12).
g   Four-vector potential $A = (\phi, c\mathbf{A})$ (Section 9.12).

9.14   What we have not discussed

In the next, and last chapter we go back to examine briefly some of the experimental evidence that was available in 1905 and which denied the existence of the ether.

Thus the end of the present chapter marks the end of our introduction to Einstein's special relativity and a few of the developments that followed. However, the theory of much of physics has been reformulated and made relativistically covariant. Some of these subjects naturally follow what we have presented but have not been discussed. They are within the reach of the interested reader. Examples include the Lagrangian formulation of relativistic particle mechanics, the energy–momentum tensors, more electromagnetism, angular momentum and spin, hydrodynamics, and thermodynamics.

Problems

9.1   Protons 1 of kinetic energy T are incident on stationary protons 2 in a laboratory. Elastic scattering occurs if a final state after a collision contains a scattered proton and a recoiling proton only. Show that the minimum angle θ_{min} between the direction of the momentum of these two final particles in the laboratory is given by

$$\cos \theta_{min} = \frac{T}{T + 4 M_p c^2}.$$

(Thus if $T \ll M_p c^2$, $\theta_{min} = 90°$.)

9.2   A particle or (a photon) has four-momentum p in an inertial frame Σ. Consider an observer moving with four-velocity $\bar{v}$ in the same frame. Show that the observer finds

1   the energy, E, of the particle (or photon) is $p \cdot \bar{v}/c$,
2   the magnitude of the particle's momentum, P, is given by

$$cP = \sqrt{(p \cdot \bar{v}/c)^2 - p \cdot p}.$$

If there are two particles with four-momenta p_1 and p_2 in Σ, show that the observer will find an angle θ_{12} between their momenta given by

$$\cos \theta_{12} = \frac{(p_1 \cdot \bar{v})(p_2 \cdot \bar{v}) - (p_1 \cdot p_2)c^2}{\sqrt{(p_1 \cdot \bar{v})^2 - p_1^2 c^2}\sqrt{(p_2 \cdot \bar{v})^2 - p_2^2 c^2}}.$$

9.3   In Example 7 of Section 9.2 we derive a formula for $\cos \theta_{12}$, where θ_{12} is the angle between $\mathbf{P}_1$ and $\mathbf{P}_2$ in the centre-of-mass of a group of n particles. Show that the formula gives $\cos \theta_{12} = -1$ if $n = 2$.

9.4   In an inertial frame Σ a particle of mass M and energy E decays into two identical particles of mass of M_a. One of the two has energy E_a and its trajectory makes an angle θ with the direction of the original particle in Σ. Show that

$$\left[(E^2 - M^2 c^4)(E_a^2 - M_a^2 c^4)\right]^{1/2} \cos \theta = E E_a - \tfrac{1}{2} M^2 c^4.$$

9.5 Very energetic electrons are scattered from protons stationary in a laboratory:

$$e^- + p \rightarrow e^- + X^+.$$

If the scattering is elastic X^+ is a proton. If the scattering is inelastic X^+ is two or more particles with a total charge of $+1$. Each incident electron has energy E_1 (four-momentum p_1). Electrons are detected at an angle θ to the direction of the incident electrons and the energy of each is measured. Consider the case in which a scattered electron has energy E_3 (four-momentum p_3) after scattering from a proton (four-momentum p_2). Energy and momentum has been transferred to the proton given by a four-momentum $q = p_1 - p_3$.

a If $Q^2 = -q \cdot q$, evaluate Q^2 in the laboratory in terms of E_1, E_3 and θ.
b Find a formula for the centre-of-mass energy W (invariant mass) of the system X in terms of invariants involving p_1, p_2 and p_3. Derive a formula for W in terms of quantities observed in the laboratory.
c Show that $Q^2 = 2(E_1 - E_3)M_p c^2$ when the scattering is elastic ($X^+ =$ proton).

9.6 Relative to an inertial frame Σ, two particles a and b have constant velocities $\mathbf{v}_a$ and $\mathbf{v}_b$. Write down the components of the four-velocities $\bar{v}_a$ and $\bar{v}_b$ of the particles in Σ.

a If Σ' is the inertial frame in which 'a' is at rest and $\mathbf{u}$ is the velocity of 'b' in this frame, write down the components of the four-velocities in Σ'. By considering an appropriate invariant, or otherwise, show that

$$u^2 = c^2 \left(1 - \frac{(1 - v_a^2/c^2)(1 - v_b^2/c^2)}{\left(1 - (\mathbf{v}_a \cdot \mathbf{v}_b/c^2)\right)^2} \right).$$

b In an inertial frame Σ'', a and b have equal and opposite velocities. By considering the form of the four-velocities of a and b in Σ'', or otherwise, find the velocity of Σ'' with respect to Σ.

9.7 The idea of the aberration of light may be widened to include the aberration of other directions of motion. Thus if a direction of motion is at an angle θ' to the z'-axis in Σ' and at an angle θ to the z-axis in Σ, we have

$$\tan \theta = \frac{\sin \theta'}{\gamma_u (\cos \theta' + \beta_u \phi)}.$$

Show that

1 $\phi = 1$ for light in a vacuum,
2 $\phi = v'_{ph}/c$ for the direction of a wave front with phase velocity $\mathbf{v}'_{ph}$, and
3 $\phi = c/v'$ for the direction of the velocity $\mathbf{v}'$ of a particle of non-zero rest mass.

9.8   An optical medium of refractive index n is stationary in an inertial frame Σ. Write down the energy and momentum components in Σ of a four-vector describing a photon of angular frequency ω moving at an angle θ with respect to the z-axis. Show that when observed in frame Σ' the angular frequency is given by

$$\omega' = \omega \frac{1 - n\beta_u \cos\theta}{\sqrt{1 - \beta_u^2}}.$$

Calculate the invariant mass (Section 3.11) of the photon and comment on its value when $n > 1$.

9.9   A medium, transparent to visible light and of refractive index n, is moving with speed u ($\ll c$) in a given direction. Visible light is transmitted through the medium in the same direction. The speed of this light observed from a stationary frame is v. Fresnel's formula for v (equation (10.8)) is

$$v = \frac{c}{n} + u\left(1 - \frac{1}{n^2}\right).$$

Show that this result may be obtained from either

1 the formula for the relativistic addition of velocities, or
2 the proper transformation of a wave four-vector as defined in equation (9.21).

9.10 A plane parallel shock wavefront travels through interstellar space in the direction of the unit vector $\mathbf{n}$ at speed v. The wavefront passes the origin at time $t = 0$. Write down an equation that must be satisfied by the coordinates $(t, \mathbf{r})$ of an event in the wavefront.

In the rest frame of the Earth two interstellar shock wave fronts intersect obliquely. In this frame the angle between the two shocks is 2θ and the wave fronts are moving with speed v. Show that if $\sin\theta < v/c$, there is an inertial frame in which the wave fronts collide head-on with equal and opposite velocities, and that in this frame the speed of propagation of one wave front is

$$w = \frac{1}{\cos\theta}\sqrt{v^2 - c^2\sin^2\theta}.$$

9.11 Show that $c\mathbf{B} \cdot \mathbf{E}$ and $\mathbf{E} \cdot \mathbf{E} - c^2\mathbf{B} \cdot \mathbf{B}$ are separately invariant under the Lorentz transformation from Σ to Σ'.

9.12 A charge q is at rest at the origin of Σ'. Therefore in Σ it moves with a speed u along the z-axis. The problem is to find the electric field $\mathbf{E}$ as a function of time at a point P in Σ not on the z-axis. The configuration of the field has cylindrical symmetry so it is convenient to choose P to be a point on the y-axis in Σ having spatial coordinates $x = 0$, $y = Y$ and $z = 0$. A measurement of $\mathbf{E}$ at P is made at time $t = T$. We call this event $\mathcal{P}$ which has, therefore, coordinates in Σ that are, in our usual notation, $(ct, \mathbf{r})_\mathcal{P} = (cT, 0, Y, 0)$.

1 Transform to find the coordinates of event $\mathcal{P}$ in Σ'.
2 The charge q is stationary at the origin of Σ'. Calculate the electric field $\mathbf{E}'$ at $\mathcal{P}$.
3 Transform $\mathbf{E}'$ from Σ' to Σ, to find $\mathbf{E}$ (equations (9.44)), expressing the result as a function of cT and Y.
4 Show that, at all times T, the electric field $\mathbf{E}$ lies i the direction from the instantaneous position of q to the point P.
5 Find the maximum value of E_y. How is it related to the maximum value of $E'_{y'}$?
6 Find an expression for the components of $\mathbf{B}$ at P in Σ as a function of T.
7 Sketch the variation of E_y, E_z, and B_x with the position of q in Σ.

9.13 In a free-electron laser a beam of relativistic electrons has a velocity v along the z-axis of a laboratory frame Σ where there is a spatially periodic magnetic field given by

$$B_y = B_0 \cos(2\pi z/L).$$

This field causes the electrons to make small transverse oscillations and to radiate electromagnetic waves. Find the frequency of oscillations in the rest frame of the electrons and then calculate the wavelength, in the laboratory frame, of the radiation emitted in the z direction.

If the electrons have energy 200 MeV and $L = 20$ mm, in what region of the electromagnetic spectrum does this radiation occur?

9.14 A rod of length slightly greater than 1 m is at rest in Σ, lying across a slot which has a width of 1.0 m.

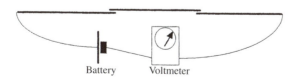

Battery   Voltmeter

The material of which the rod and both sides of the slot are made is metallic. If the rod is at rest and making good contacts it completes an electrical circuit in which the slot is the only gap in the absence of the rod.

The rod moves slowly from left to right (in the figure) and the electric circuit is completed for a short time. (A good electrical contact is assured at an end of the rod which overlaps with the slot.) If the rod moves at $v/c = 0.8$, again from left to right, it would appear that the contraction of the rod would prevent any completion of the electric circuit. If the slot moves from right to left with $v/c = 0.8$, the slot would contract and it would appear that the circuit would be completed for a period of time. Whether completion of the circuit would occur or not, appears to depend on the inertial frame in which the events are observed. This cannot be the case. Can you resolve this paradox?

9.15   The transformation of the components of a four-vector or of the components of the interval from inertial frame Σ to Σ' is given by

$$dx'^{\alpha} = \Lambda^{\alpha}_{\mu} dx^{\mu}.$$

Find the elements of Λ (and of $\bar{\Lambda}$, equation (9.64)) for the standard arrangement of Σ and Σ'.

Consider the metric tensor $g^{\mu\nu}$. Transform it from Σ to Σ' and thereby show that it is invariant.

9.16   Given the results of Problem 7.16 and the transformations given in equations (9.44), derive equations (5.15) from equation (5.14).

9.17 R and S are mixed tensors with dimension 4 and rank-5 and 2 respectively. Show that P is also a tensor where

$$P^{\lambda\kappa\eta}_{\beta\gamma} = R^{\lambda\kappa}_{\alpha\beta\gamma} S^{\alpha\eta}.$$

9.18   Transform the covariant tensor $F_{\mu\nu}$ (equation (9.80)) from Σ to Σ'. Show that this gives the same results for the fields $\mathbf{E}'$ and $\mathbf{B}'$ in Σ' in terms of the fields $\mathbf{E}$ and $\mathbf{B}$ in Σ as obtained in Section 9.8 (equations (9.44)). Use the results of Problem 9.15. (Take particular care about the sign of β in transforming from Σ to Σ' the components of a covariant tensor.)

9.19   Verify that equations (9.80)–(9.82), that is

$$\partial^{\mu} F_{\mu\nu} = \sqrt{\frac{\mu_0}{\varepsilon_o}} J_{\nu}, \quad \partial_{\lambda} F_{\mu\nu} + \partial_{\mu} F_{\nu\lambda} + \partial_{\nu} F_{\lambda\mu} = 0,$$

$$\text{and} \quad \partial^{\nu} \partial^{\mu} F_{\mu\nu} = \sqrt{\frac{\mu_0}{\varepsilon_o}} \partial^{\nu} J_{\nu} = 0,$$

give Maxwell's equations when $F_{\mu\nu}$ is the tensor with components given in equation (9.79).

9.20 The quantity $\varepsilon^{\kappa\lambda\mu\nu}$ is a rank-4 contravariant tensor with the same components in all inertial frames:

$\quad$ 0 $\quad$ if any two or more indices are the same,

$\quad$ +1 $\quad$ if $\kappa\lambda\mu\nu$ is an even number of permutations of 0123,

$\quad$ −1 $\quad$ if $\kappa\lambda\mu\nu$ is an odd number of permutations of 0123.

A permutation is the interchange of any pair of neighbouring indices. Thus, for example, $\varepsilon^{0132} = -1$ and $\varepsilon^{1032} = +1$. Now consider the equation

$$\varepsilon^{\kappa\lambda\mu\nu}\partial_\lambda F_{\mu\nu} = 0.$$

1 What is the rank of the left-hand side and how many independent equations are there in this equality?
2 Is this equality Lorentz covariant? Explain your answer.
3 Show that this equality is the same as equation (9.81).

10 A postponed prelude

Introduction

This chapter has the form of a postponed prelude for special relativity. A prelude because it describes some of the experimental and theoretical work done before 1905. Postponed because these subjects might properly have appeared in a first chapter, although we chose not to do that. So here they are: Sections 10.2–10.6 on the experiments relevant to finding the ether and the theories that tried to understand the failure of these experiments while remaining true to Galileo and to Maxwell.

But the voices of the exponents of the ether were not instantly extinguished in 1905. We briefly review the post-1905 story in Section 10.7.

**10.2 Maxwell and the ether

Maxwell in the mid-nineteenth century unified electricity and magnetism. He found that electromagnetic disturbances travelled in a vacuum with a speed c that, in modern SI units, is equal to $1/\sqrt{\varepsilon_0\mu_0}$. This number was close to the then known speed of light and Maxwell therefore identified light as an electromagnetic wave. He thereby added optics to the subjects unified in his theory. However, as we have found in Section 5.4, Maxwell's equations were not consistent with Galilean relativity.

Physicists were familiar with various kinds of waves but in all cases the transmission of a wave required a medium and the wave was associated with varying displacements of the medium. Before Maxwell the waves of light were supposedly carried by the so-called *luminiferous ether*. Another ether supposedly existed as the medium which transmitted the forces between charges and between currents. What we know as an electric field was considered to be a stress in this ether, and the energy of the field was associated with the stress and the strain in the ether. After Maxwell, this ether had the duty of acting as the medium of transmission of electromagnetic waves, including light. In addition, Maxwell's equations were expected to be correct only in a frame of reference stationary with respect to the ether. Of course the ether had to have some unusual properties: it was without mass and the Earth appeared to be transparent to it.

Without worrying about its properties we may think of this ether as defining a preferred inertial frame (the ether frame) in which Maxwell's equations were correct and the speed of light was c. Galilean relativity continued to apply and the velocity of light on Earth would depend on the velocity of the Earth relative to the ether.

This motion of the Earth through the ether was called *the ether drift*. Suppose the ether frame is stationary with respect to the Sun. Then the Earth will have speed, v, in the ether frame close to 30 km s^{-1}. In this pre-Einstein world view, the speed of light relative to the Earth in a direction fixed on the Earth but parallel to the plane of the Earth's rotation around the Sun (the ecliptic) would vary between approximately $c+v$ and $c-v$ during the course of one day. Approximately because this neglects the angle between the axis of rotation of the Earth and the normal to the ecliptic, and the small effect of the speed of the surface of the Earth in its daily rotation. However, the speed could only be measured by timing over a path to a distant mirror and back. In such 'there-and-back' experiments the fractional variations in speed that are first order in v/c cancel (equation (10.1)) and the measured speed changes will only depend on $(v/c)^2$. This is 10^{-8} for $v = 30$ km s^{-1}. The sensitivity required forced hunters of the ether to interference methods for detecting changes in the speed of light, and to other methods of ether detection.

**10.3   The hunt for the ether: Michelson and Morley

A. A. Michelson invented an optical interferometer and used it in 1881 to search for the ether; specifically to detect the effect of the motion of the Earth around the Sun as it carried the interferometer through the ether. The basic elements of the device are shown in Figure 10.1. Michelson's 1881 attempt was insufficiently sensitive to detect with certainty the expected effects and he joined E. W. Morley in constructing an interferometer with a twenty-fold increase in sensitivity. The results were published in 1887.

Their apparatus was essentially as in Figure 10.1 with arrangements that allowed the whole to be rotated around a vertical axis. In addition, each arm of the interferometer was increased to a length of 11 m by folding that distance into a space of about 2 m with mirrors (see Problem 10.1). These few words do not do justice to the careful design and planning necessary to ensure that the optics were stable for long periods of time and free from disturbance from the rotation or external factors.

To see how this method works let us suppose that the velocity through the ether is of magnitude v and in the direction of M–M1 (Figure 10.1). The time taken for the light to travel from M to M1 and back to M is, in the ether model,

$$t_1 = \frac{l_1}{c-v} + \frac{l_1}{c+v} = 2l_1\left(\frac{c}{c^2-v^2}\right). \tag{10.1}$$

The light journey from M to M2 and back to M is along two equal sides of a triangle having a base that is the distance moved by M along the direction towards M1, so that the time taken for this trip satisfies

$$t_2 = \frac{2}{c}\left(l_2^2 + \left(\frac{v\,t_2}{2}\right)^2\right)^{1/2}.$$

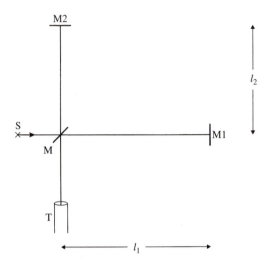

Figure 10.1 The Michelson interferometer. The operation is most easily described
if we assume a monochromatic light source, S. Light from S is divided
into two beams by division of amplitude at a half-silvered mirror M. One
beam travels a distance l_1 to mirror M1 where it is reflected and returns
to M where a fraction is reflected into a viewing telescope T. The second
beam travels a distance l_2 to mirror M2, is reflected and returns to M
where a fraction continues to T. Thus the two beams are recombined at
M and interfere constructively or destructively in the field of view of the
telescope. By slightly tilting M1 and M2 from being mutually perpendicular, parallel interference fringes may be observed in the telescope.
If l_1 or l_2 is gradually changed the fringe pattern will move one way or
the other. Each change of path difference between the two beams of one
wavelength will replace each bright fringe by a next neighbour bright
fringe. In Section 10.2 we shall call this a (complete) fringe shift of one.

   In describing this interferometer we have not dwelt upon the details of
the optics involved or of the steps that must be taken to obtain high quality
fringes. Readers who are doing a course of study involving experimental
physics have surely used or seen such an interferometer, or will do so.

Hence

$$t_2 = \frac{2\,l_2}{\sqrt{c^2 - v^2}}.$$ (10.2)

The amount by which the path time via M1 exceeds that via M2 is

$$\Delta t = t_1 - t_2 = \frac{2}{c}\left(\frac{c^2}{c^2 - v^2}l_1 - \frac{c}{\sqrt{c^2 - v^2}}l_2\right).$$ (10.3)

If the apparatus is rotated through 90° so that the path M–M2 is now where M–M1 was previously then the time via M1 exceeds the time via M2 by

$$\Delta t' = t_1' - t_2' = \frac{2}{c}\left(\frac{c}{\sqrt{c^2 - v^2}}\,l_1 - \frac{c^2}{c^2 - v^2}\,l_2\right). \tag{10.4}$$

If the light has a wavelength λ, the difference $\Delta t' - \Delta t$ becomes a change in phase between the two beams interfering at the telescope of $c(\Delta t' - \Delta t)2\pi/\lambda$. For each 2π the fringe pattern will shift by one complete fringe so that the fringe shift in number on rotating 90° is

$$n = c(\Delta t' - \Delta t)/\lambda = 2\frac{l_1 + l_2}{\lambda}\left(\frac{c}{\sqrt{c^2 - v^2}} - \frac{c^2}{c^2 - v^2}\right). \tag{10.5}$$

Expanding and dropping terms in powers greater than $(v/c)^2$ gives

$$n = -\frac{l_1 + l_2}{\lambda}\left(\frac{v}{c}\right)^2. \tag{10.6}$$

Michelson and Morley had $l_1 = l_2 = 11$ m and they used light of wavelength 589 nm. If the ether was stationary with respect to the Sun then v was the Earth's speed around the Sun, 29.8 km s^{-1}, and the proposed rotation should give $n = 0.37$.

In practice, Michelson and Morley could not know the velocity of the Earth through any supposed ether. They measured the fringe pattern position with respect to a fiducial marker in the telescope every 22.5° as the interferometer was steadily rotated through 36 revolutions. This was done six times in the course of three days and nights during the month of July, 1887. They had intended to repeat such measurements at three month intervals throughout a year to be sure that the relative movement of the Earth and of the ether was not accidentally small during their July measurements. But they did not.

Michelson and Morley concluded that the ether drift (in July 1887) was less than a quarter of the Earth's orbital speed. Their experiment is justly famous as the first, and almost the last, sensitive attempt to detect the ether before Einstein's 1905 paper. Although their result is not a definitive null result it was consistent with zero ether drift and far from the result expected in the ether theory. Lorentz constructed a theory (see Section 10.5) assuming a null result. The experiment was repeated with increasing precision and under different conditions from 1902 to 1930. Panofsky and Phillips[1] have an accessible review reproduced from the scientific literature. The observations from all experiments upto 1930 were subjected to the same analysis and none has a result that is not consistent with zero ether drift. Since 1930 techniques have improved and the experiment has been repeated at increased sensitivity. A method using lasers is described in the book by Hecht and Zajac[2] (1964, giving $v < 30$ m s^{-1}).

1 Panofsky, W. K. H. and Phillips, M., *Classical electricity and magnetism*, 2nd edn, Reading, Mass.: Addison–Wesley, 1962, p. 277.
2 Hecht, E. and Zajac, A., *Optics*, Reading, Mass.: Addison–Wesley, 1974, p. 322.

**10.4 The hunt for the ether: Trouton and Noble, Rayleigh, and Brace

Consider two equal and opposite electric charges fixed at either end of a rod that is stationary on the Earth. There is a Coulomb attraction between the charges which acts along the rod. Suppose the ether exists and that it is ether frame in which Maxwell's equations are correct. Then the motion of the Earth through the ether means that the two charges constitute two currents which generate magnetic fields. The magnetic field due to one charge exerts a force on the other charge and the other way round. These two forces are equal and opposite but not generally collinear. Therefore there can be a couple acting on the rod. This couple is zero if the rod is parallel to or perpendicular to the velocity of the Earth through the ether. In between these positions the couple tends to turn the rod to bring the line between the charges perpendicular to this velocity. The rod parallel to the velocity is a position of unstable equilibrium.[3]

In 1903 F. Trouton and H. R. Noble attempted to detect such a couple using a parallel plate capacitor suspended by a torsion wire so that the plates of the capacitor were in a vertical plane. The capacitor could be charged. The motion through the ether of the charges resident on the plates would produce a couple tending to turn the capacitor so that the plates would become parallel to the velocity. Any couple would be detected by a rotation of the capacitor on its suspension if the capacitor was not oriented for zero couple. The sensitivity was easily sufficient to detect the couple expected due to the Earth's velocity through the ether equal to its speed in orbit around the Sun. No couple was detected.

Another approach to detecting the ether involved attempting to find the effect on isotropic, transparent materials moving through it. The expected effect is that the dielectric constant, which was normally isotropic, would become anisotropic. For light this translates into a refractive index that depends on the polarisation and direction of light in the material. This property exists normally in anisotropic crystals and is called *birefringence*.[4] Experiments by Lord Rayleigh in 1902 and separately H. B. Brace in 1904 failed to find the expected birefringence.

**10.5 Saving the ether with Lorentz!

Lorentz had known about Michelson's first attempt to find the ether and had found an error in his analysis (although not the first to notice it). The later result of the Michelson–Morley collaboration must have been a puzzling matter for the nineteenth century physicists who relied upon the existence of the ether and were aware of this development. FitzGerald (see Chapter 1) and Lorentz separately proposed a solution. It was that moving bodies contracted along a direction parallel to their velocity $\mathbf{v}$ though the ether by a factor $\sqrt{1 - v^2/c^2}$. This is the Lorentz–FitzGerald contraction.

---

3 Panofsky and Phillips, *op. cit.*, pp. 274–5.
4 Hecht and Zajac, *op. cit.*, pp. 235–7.

Look at equations (10.3) and (10.4) to see how this affects the Michelson–Morley experiment. In equation (10.3) the length of the arm of the interferometer in the direction of the supposed velocity through the ether is l_1. If it contracts to $l_1\sqrt{1 - v^2/c^2}$ then

$$c\Delta t = \frac{2(l_1 - l_2)}{\sqrt{1 - v^2/c^2}}. \tag{10.7}$$

Similarly, for the 90° position it is l_2 in equation (10.4) that contracts by the same factor. That means $\Delta t' = \Delta t$ and there is no fringe shift on rotating the interferometer through 90°. Therefore the proposed contraction explains the null result from Michelson and Morley.

Between 1892 and 1905 Lorentz constructed and developed an electrodynamic theory based on this hypothesis of a longitudinal contraction of bodies moving through the ether.[5] The forces within and between molecules constituting such a body were assumed to be electromagnetic in origin and subject to this contraction. The size of the electrons, which were known to be constituents of all molecules, was included in this contraction. Thus the longitudinal dimension of a body contracted as a consequence of a change in the size and equilibrium separation of the molecules. With this theory Lorentz was able to explain the null result of the searches for the ether completed by the time of Einstein's first paper on relativity. In addition, it gave the same formula for the increase in mass with speed of the electron as given by Einstein in 1905 (see Section 4.3).

Although long after 1905 and not a part of the prelude to special relativity, R. J. Kennedy and E. M. Thorndike in 1932 repeated the Michelson–Morley experiment with an interferometer having arms of unequal length. Their result showed that the ether with the property of causing Lorentz–FitzGerald contraction was not compatible with experimental observation. To see how it works consider equation (10.7) which gives the difference in time for light to travel there and back along the two arms in the presence of Lorentz–FitzGerald contraction:

$$\Delta t = t_1 - t_2 = \frac{2(l_1 - l_2)}{\sqrt{c^2 - v^2}},$$

where v is the ether drift speed taken to be in the direction of arm 1. As we have seen, if the interferometer is rotated through 90° there could be a new time difference $\Delta t'$. However in this case with contraction $\Delta t = \Delta t'$ and no fringe shift occurs on rotation. And, of course, Michelson and Morley had $l_1 = l_2$ so that $\Delta t = \Delta t' = 0$ whatever the orientation of the interferometer. If $l_1 \neq l_2$ but v

5 Lorentz, H. A., Two sections in *The principle of relativity*, an English translation of a collection of original papers on the special and general theories of relativity by Lorentz, Einstein, Minkowski, and Weyl. It includes Einstein's two 1905 papers on relativity. New York: Dover Publications, 1952.

changes to v' then the time difference Δt changes to $\Delta t'$ where

$$\Delta t' - \Delta t = 2\frac{(l_1 - l_2)}{c}\left(\frac{c}{\sqrt{c^2 - v'^2}} - \frac{c}{\sqrt{c^2 - v^2}}\right)$$

and there is a fringe shift:

$$n = c\frac{(\Delta t' - \Delta t)}{\lambda} = \frac{(l_1 - l_2)}{\lambda}\left(\left(\frac{v'}{c}\right)^2 - \left(\frac{v}{c}\right)^2\right).$$

So Kennedy and Thorndike set out to look for speed variations. The interferometer was fixed and they measured the fringe shifts during many 24-hour periods and again at several periods during a year. The speed could have been a vector sum of the effect of the Earth's daily rotation, the orbital speed around the Sun, and the speed of the Sun through the ether. This sum is changing daily and annually. Kennedy and Thorndike detected no change in speed greater than about $10\,\text{km}\,\text{s}^{-1}$ and their result was consistent with no change. Ether with contraction was not successful.

It is worth noting that where Michelson and Morley had $l_1 = l_2 = 11\,\text{m}$, Kennedy and Thorndike were restricted to $l_1 - l_2 = 16\,\text{cm}$ by the coherence length of the light from their source. This meant that their method of measurement of fringe shift had to be very precise and the interferometer very well isolated from extraneous effects.

**10.6   Dragging the ether

Another attempt to save the ether was the hypothesis that the ether was attached to all bodies with mass and therefore dragged along by such bodies (*ether drag*). This explains the null results obtained by Michelson and Morley, by Trouton and Noble, by Rayleigh, and by Brace. However, there were two pieces of experimental evidence that were in contradiction to this hypothesis

1   The existence of stellar aberration has been discussed in Section 6.5. If the ether was dragged with the Earth and presumably by the air and by the structure of a telescope then there would be no stellar aberration, contrary to observation.

2   The phase velocity of light in a transparent medium is c/n, where n is the refractive index of the medium. If the medium is moving with respect to an observer with a speed u then the ether drag would give the light (moving in the same direction) a speed $c/n + u$ with respect to the observer. Fizeau in 1851 had measured how the speed of light changed with the speed of moving water through which the light is sent. His apparatus is described briefly in Figure 10.2 and it depends on the interference between beams of coherent light sent in opposite directions through flowing water. The path length was $2.86\,\text{m}$ and the water flow was varied between 4 and $7\,\text{m}\,\text{s}^{-1}$. At maximum speed he observed a shift of 0.46 fringes when the flow of water was reversed, where 0.96 was expected with complete ether drag. The observed shift was consistent with a formula previously derived by Fresnel. For light moving

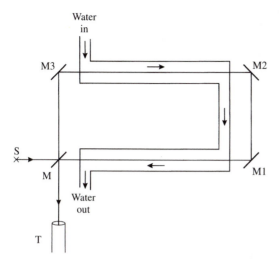

Figure 10.2 Fizeau's apparatus for measuring the speed of light in moving water. Light from a monochromatic source S is divided into two beams by the half-silvered mirror M. One beam traverses the path from M to mirror M1 to mirror M2 to mirror M3 and back to M. The other beam traverses the path from M to M3 to M2 to M1 to M. A fraction of the two beams are recombined at M so as to travel into the viewing telescope T where their interference is observed. A suitable vessel conducts a stream of water along the path indicated. One light beam passes through the water against the flow, the other with the flow. With no flow there is no optical path difference and the light at the centre of the field of view of T will interfere constructively. Flow was expected to change the phase velocity of the light in the water relative to the laboratory. The effect would be opposite for the two beams and an optical path difference would be established. The optical path difference would change the interference condition. If the optics is arranged to give parallel fringes, this fringe pattern will move as the flow is changed. The fringe shifts allowed Fizeau to determine the phase velocity as a function of the speed of the water flow.

with the flow that has speed u, the speed v measured by a stationary observer is given by

$$v = \frac{c}{n} + u \left(1 - \frac{1}{n^2} \right).$$

(10.8)

Fizeau repeated the experiment with air flowing at 25 m s^{-1} but could observe no fringe shift, again consistent with Fresnel's formula. These results were not in accord with ether drag. Problem 9.9 asks the reader to prove equation (10.8).

Of course, there were proposals for a partial ether drag depending on refractive index and therefore on wavelength. But constructing an ether theory that was consistent with that and with all other optical phenomena, required too many artificial hypotheses to be satisfactory.

**10.7   1905 and after

The year 1905 was the watershed when Einstein's alternative to ether theories became available. Table 10.1 gives a simple list of the pre-1905 ether theories and the experiments that pointed to their demise. Note that the ether with the Lorentz–FitzGerald contraction was not confronted by an optical experiment until 1932.

Einstein's postulates mean that the speed of light is the same in all directions and in all inertial frames and that there is no preferred ether frame, or no ether, if that is different. Thus all the experiments that we have described, or related ones that we have not the space–time to mention, were bound to give a result consistent with zero ether drift. Fizeau's result may be explained by the formula for the relativistic addition of velocities (see Problem 9.9).

However, the ether plus Galilean relativity did not lie down and die instantly. For several decades after 1905 there were two themes:

1   Continuing attempts to sustain the ether or to find alternative theories that retained Galilean relativity.
2   To accumulate experimental evidence for the correctness of special relativity and to refine theoretically its logical and mathematical foundations.

Of course, these themes are interwoven: for example, more sensitive searches for the ether reinforced the case for special relativity.

For theme one we can briefly mention two theories that were current in the decade after 1905 and that retained Galilean relativity.

1   After 1905 Lorentz knew about Einstein's relativity but held to his belief in the existence of the ether. As we have seen, before 1905 he had incorporated into his theory, and justified, the hypothesis of the contraction of bodies moving through the ether. After 1905 he borrowed a result from Einstein adding to his theory the hypothesis that moving clocks run slow relative to clocks stationary

Table 10.1 The pre-1905 ether theories and the experimental results that falsified them

Theory	Falsified by
Stationary ether without contraction	Michelson and Morley, Trouton and Noble, Lord Rayleigh, and Brace
Ether attached to bodies with mass	Stellar aberration and Fizeau
Stationary ether with Lorentz–FitzGerald contraction	Kennedy and Thorndike (but not until 1932)

in the ether. He appeared to hope that he had deduced from the equations of electromagnetism what Einstein had postulated. This theory made the same predictions as special relativity and, therefore, on the available experimental results had equal claim to validity. To many, Einstein's replacement of the contraction hypotheses with the postulates of special relativity was not necessarily an acceptable exchange. But Lorentz's theory had none of the fundamental simplicity of special relativity, and, with the clarity of hindsight, we know that Lorentz's theory was going nowhere. The future would vindicate special relativity. That should not be taken to belittle Lorentz's efforts; he was a giant and his name rightly graces the transformations of special relativity.

2   The so-called *emission theories*. They are based on a reformulation of Maxwell's equations and have in common the hypothesis that the speed of light is always c/n relative to the source, where n is the refractive index of the medium in which the speed is observed. These theories were in accord with stellar aberration, Fizeau, and Michelson and Morley. The final nails in their common coffin were hammered in by a measurement of the speed of photons from the decay of very energetic neutral pions (see Table 10.2).

Table 10.2 A selection of observations that are relevant to the verification of the postulates of special relativity and of the Lorentz transformations. References are made to other sections, to text books that give some explanation, or to the original publication if nothing more accessible is available. The precision is only meant as a rough guide and may apply only to a restricted range of possible conditions

Ether drift interpretation			Precision
Observation	v	*Meaning for relativity*	
Michelson and Morley[1] and a successor[2]	$<8\,\mathrm{km\ s^{-1}}$ $<30\,\mathrm{m\ s^{-1}}$	The two-way speed of light is the same in all directions	1 in 4×10^5 (1887) 1 in 1×10^7 (1964)
Kennedy and Thorndike[3]	$<10\,\mathrm{km\ s^{-1}}$	The speed of light is the same in all inertial frames	1 in 3×10^4 (1932)
Speed of photons from decay of energetic π^0 mesons[4]	—	The speed of light is independent of the velocity of the source	1 in 4×10^5 (1964)
Dispersion of signals from pulsar over 13 decades of frequency[5]	—	The speed of light is independent of frequency	1 in 10^{14} (1972)
Speed of 11 GeV electrons against speed of photons[6]	—	Existence of a limiting speed	1 in 2×10^5 (1973)
Rate of decay of muons[7]	—	Confirmation of time dilation	1 in 2×10^3 (1977)

1 See Section 10.3.
2 See Section 10.3, about a laser experiment described in the textbook by Hecht and Zajac.
3 See Section 10.5.
4 Jackson, J. D. *Classical electrodynamics*, 2nd edn, New York: John Wiley & Sons, 1975, pp. 512–4.
5 Jackson, *op. cit.*, p. 515.
6 Jackson, *op. cit.*, p. 531.
7 See Section 7.11, Bailey, J., *et al.*, *Nature* **268**, 1977, p. 301.

For the second theme, we have given in Table 10.2 a selection of observations that are relevant to the verification of (1) the postulates of special relativity as applied to light and electromagnetic waves and of (2) a couple of the more immediate consequences of the Lorentz transformations. But this theme does not end with those subjects. Chapters 2 and 3 are concerned with the properties of relativistic energy and momentum, and in particularly their conservation. The success of relativistic kinematics at energies where the particle speeds are close to that of light is taken for granted but it adds validity to special relativity.

Soon after 1905 Einstein moved on to what was to become general relativity. As far as the second theme is concerned, it was Minkowski who took the first steps in refining the mathematical foundations of relativity (1908). He emphasised the four-dimensional nature of space-time as against $3 + 1$ space and time (see Section 7.17). Einstein by using tensors in general relativity stimulated their use in special relativity (see Sections 9.11 and 9.12), thereby allowing the proper implementation of Minkowski's ideas. Thereafter, much classical physics was recast in relativistic terms; electrodynamics, elasticity, hydrodynamics, thermodynamics, to name a few. Further important developments came when relativity was wedded to quantum mechanics. We have briefly looked at the discovery of the Dirac equation. Since then relativistic Quantum Field Theory (QFT) has told us how to quantise fields and how to deal with fields interacting. That theory is a fundamental ingredient of our understanding of quantum processes. Such processes are observed and the theory which correctly describes them could not have been developed in the absence of special relativity. One offshoot is the theory of the electromagnetic and weak interactions of the fundamental fermions and bosons. In its range of applicability and in the experimentally confirmed accuracy of its predictions it is the most successful physical theory that we have. Another offshoot is quantum chromodynamics: this is the theory of the interaction of the quarks and gluons which are the constituents of the nucleons and related particles. Although believed to be correct, it is very resistant to accurate calculations. So, although we may not have to worry about its effect when v/c is small, special relativity is everywhere; even in the presence of gravity or of accelerations, it applies locally.

It is not easy to trace and spell out the logical development of the subject before Einstein. The student reader may require an easy-to-remember list that is not encumbered by the quibbles of controversy or the fog of false trails. Table 10.1 presents the pre-Einstein experimental facts that, seen in retrospect, pointed to the demise of the ether and to the birth of special relativity.

At a time nearly a century after Einstein's first paper on relativity, it would appear there is no more reason to spell out special relativity's successes than there would be to do the same for Newton's mechanics. Each is, in its own way, an unquestioned part of the rigorous scientific subject that we know as physics. However, we have seen that nearly half a century passed before those who doubted Einstein fell silent, at least in the scientific literature. So in Table 10.2 we have listed some of the up-to-date and precise experiments that were performed to test relativity as technical improvements made each worth while.

Problem answers

2.2 1038.3 MeV; 444.6 MeV/c; 0.4282c; 10938 MeV; 10898 MeV/c; 0.9963c.

2.3 0.86603c.

2.4 (a) 1–1.13 × 10^{-5}; (b) 6.45 × 10^{-12}.

2.5 4.40 × 10^9 kg s^{-1}.

2.6 (a) 17.34 MeV; (b) 39.27 MeV; (c) 17.84 MeV; (d) 258 MeV/c.

2.7 3.25 eV; 3.25 eV.

2.8 4.7 × 10^{-12} kg.

3.1 π: 32.4 MeV; 100.5 MeV/c; 0.5845c. p: 5.3 MeV; 100.5 MeV/c; 0.1066c.

3.2 48 MeV.

3.3 1103.8 MeV/c^2.

3.4 $\sqrt{21}\,M$; 0.646c.

3.7 (a) 1671.7 MeV; (b) 0.5133c; (c) 561.3 MeV/c.

3.8 289.6 MeV; 768.0 MeV.

3.9 0.9142c; 3494 MeV.

3.11 1.1 × 10^{15} eV.

4.1 0.192 MeV; 0.483 MeV; (a) 2.06 × 10^6 V m^{-1}; (b) 515.3 V; (c) 1.06 cm.

4.3 (a) 2.93 m; (b) 22.35 MHz; (c) 12.06 MHz.

4.4 (1) 27.2 GeV/c^2; 41.5 GeV/c^2; (2) 47.2 GeV/c^2; (3) 312.8 GeV/c^2.

4.5 (a) 4.23 × 10^7 GeV; (b) (i) 5.24 × 10^4 GeV; (ii) 9.63 × 10^7 GeV.

6.5 (1) 547.7 nm; (2) $c/11$.

6.6 0.2422c; 406.3 nm.

6.8 0.6c; 65.4°.

6.9 Hydra is receding at angle θ to the line of sight with speed βc so that

$$\gamma(1 + \beta \cos\theta) = 475/394.$$

6.10 $2\cos^{-1}(v/c)$.

6.11 In the first two cases the plane containing the first photon and the z-axis also contains the second photon but always on the other side of that axis. (i) 154.8 MeV. The second photon has the same energy and is moving at the same angle with respect to the z-axis. (ii) 29.4 MeV; 280.2 MeV; 6.03°. (iii) 294.1 MeV; 15.5 MeV; directly backwards.

6.13 (1) 166 MeV/c; (2) 344 MeV/c; 29°.

6.14 191.8 MeV; 10.7°; (1) 11.4°; 177.7 MeV; (2) 149.1 MeV; 0°.

6.15 (1) 62.38 GeV; (2) 0.1287c in a direction in the plane of the intersecting beams and at 97.4° to both. 303.1 MeV/c at 81.85° with respect to the direction of motion of the centre-of-mass and at equal angles with respect to the directions of the incident protons.

6.16 9.03 GeV; 12.13 GeV; 5.93 GeV/c.

6.17 3.22 GeV.

6.19 47.3 MeV; 7.7 MeV.

7.4 27.9 cm; 9.31 × 10^{-10} s; 3.5%.

7.7 (1) 341 MeV/c; (2) 29.6 μm; 2572 MeV/c; 3354 MeV/c; 228 μm; 297 μm.

7.8 (a) $\tan^{-1}[Pc/\{\gamma_u(Pc - E\beta_u\sqrt{2})\}]$, where $E = \sqrt{P^2c^2 + M^2c^4}$; (b) $\tan^{-1}\gamma_u$.

7.9   6.46 T.

7.10 d/u; $vd/u(v - u)$; $d/u + d\sqrt{(1 - v^2/c^2)}/(v - u)$.

7.11 $4c/5$; $f_0/\sqrt{3}$; $f_0/3$.

7.12   456 nm.

8.8   41 years and 6 weeks old; 353 days and 17 h.

8.9   (2) $c\sqrt{1 - 1/[\gamma_u^2 \gamma_v^2 (1 - \boldsymbol{\beta}_u \cdot \boldsymbol{\beta}_v)]^2}$.

9.5   (a) $Q^2 = 4E_1 E_3 \sin^2 \theta/2$, if E_1, $E_3 \gg M_e c^2$.
       (b) $W^2 = (M_p c^2)^2 + 2(E_1 - E_3)M_p c^2 - Q^2$.

9.6   (b) $c\{\gamma_a \boldsymbol{\beta}_a + \gamma_b \boldsymbol{\beta}_b\}/(\gamma_a + \gamma_b)$.

9.8 $(\hbar\omega/c^2)\sqrt{1 - n^2}$; fractions of the energy and momentum of the photon incident on the medium are shared with the medium so that the photon alone does not have $E = Pc$.

9.10 $\mathbf{r} \cdot \mathbf{n} = vt$.

9.12   (3) $E_x = 0$, $E_y = (q/4\pi\varepsilon_0)(\gamma_u Y)/[Y^2 + (\gamma_u \beta_u cT)^2]^{3/2}$;
       $$E_z = -(q/4\pi\varepsilon_0)(\gamma_u \beta_u cT)/[Y^2 + (\gamma_u \beta_u cT)^2]^{3/2}.$$
       (5) $E_{y\,max} = (q/4\pi\varepsilon_0)[\gamma_u/Y^2] = \gamma_u(E'_{y'}$ at $y' = y)$.
       (6) $cB_x = -(q/4\pi\varepsilon_0)(\beta_u \gamma_u Y)/[Y^2 + (\gamma_u \beta_u cT)^2]^{3/2}$; $B_y = 0$; $B_z = 0$.

9.13 5.83×10^{12} Hz; 65.2 nm; extreme ultraviolet.

9.20   (1) 1; 4.

Bibliography

Internet site

http://www.aip.org/history/einstein. This is a section of the web site of the American Institute of Physics devoted to the history of the subject and in particular to Einstein. There is an extensive bibliography and links to many other sites with material about him.

Books on special relativity

Einstein, A. (2001) *Relativity: The special and general theory* (15th edn reprint with revisions), (trans. R. W. Lawson). London: Routledge.
[A translation (1920) of Einstein's book of 1916. A simple and elegant account with very little mathematics].

Einstein, A. (1956) *The meaning of relativity* (6th edn revised), (trans. E. P. Adams *et al.*). London: Methuen.
[The Stafford Little lectures on the special and general theories of relativity delivered in May 1921 at Princeton University. Appendices later added to the text. A mathematical approach].

Rindler, W. (1991) *Introduction to special relativity* (2nd edn), Oxford: Oxford University Press.
[A good basic text covering material for an advanced undergraduate course].

French, A. P. (1968) *Special relativity*, The MIT Introductory Physics Series, Van Nostrand Reinhold.
[An American text designed for university undergraduates and having a scope comparable to that of the present text].

Rosser, W. G. V. (1991) *Introductory special relativity*, London: Taylor and Francis.

Adams, S. (1997) *Relativity. An introduction to spacetime physics*, London: Taylor and Francis.
[The author aims to provide an accessible and informal route to the theory of relativity without neglecting the essential mathematics. A significant part of this book is devoted to general relativity and cosmology. The intended audience is interested senior school pupils and first year university undergraduates].

Taylor, E. F. and Wheeler, J. A. (1992) *Spacetime physics* (2nd edn), London: W. H. Freeman.
[A stimulating approach to special relativity intended for university undergraduates].

Sartori, L. (1996) *Understanding relativity: A simplified approach to Einstein's theories*, Berkeley: University of California Press.
    [A book for those with little mathematical preparation. A thorough discussion of many aspects. Last third on general relativity].
Sard, R. D. (1970) *Relativistic mechanics*, New York: Benjamin.
    [A detailed discussion of many aspects of special relativity].
Jackson, J. D. (1998) *Classical electrodynamics* (3rd edn), New York: Wiley, Chapter 11.
Schwartz, J. and McGuinness, M. (1999) *Introducing Einstein*, Cambridge: Icon Press.
    [A cartoon book that weaves together the stories of the science and of the life of Einstein, and of his development of special relativity. An easy and enjoyable read].
Epstein, L. C. (1985) *Relativity visualized*, San Francisco: Insight Press.
    [Special and general relativity explained in words and pictures, and served with humour].

A book for those who would like, after Chapter 3, to know more about the elementary particles and their properties there is:

Martin, B. R. and Shaw, G. (1997) *Particle physics* (2nd edn), Chichester: Wiley.

Translations of Einstein's two relativity papers of 1905 may be found in the following two books:

Lorentz, H. A., Einstein, A., Minkowski, H. and Weyl, H. (1952) *The principle of relativity* (trans. W. Perrett and G. B. Jeffery). New York: Dover Publications, Inc.
    [In addition to the two 1905 papers this volume includes five papers dating from 1911 to 1919 by Einstein on general relativity, and others by the authors named].
Stachel, J. (ed.) (1998) *Einstein's miraculous year*, New Jersey: Princeton University Press.
    [A commentary on the 'miraculous year' (1905) and modern (1998) translations (T. Lipscombe *et al.*) of 'the five papers that changed the face of physics'].

Beyond special relativity

The books that will lead the reader from special into general relativity are the two by Einstein and those by Epstein, Sartori, and Adams. Recent books that are more specific are as follows.

Schutz, B. F. (1992) *A first course in general relativity*, Cambridge: Cambridge University Press.
d'Inverno, R. (1996) *Introducing Einstein's relativity*, Oxford: Oxford University Press.
Wheeler, J. A. (1990) *A journey into gravity and spacetime*, New York: Scientific American Library.
    [A very readable description of the meaning of general relativity aimed at the intelligent who are without any mathematical pretensions].

Biographical works

Bernstein, J. (1997) *Albert Einstein and the frontiers of physics*, Oxford: Oxford University Press.

[One of a series that aim to combine intelligible scientific information with personal stories to portray some great scientists whose work has shaped our understanding of the natural world].

Pais, A. (1982) *'Subtle is the Lord...'. The science and life of Albert Einstein*, Oxford: Oxford University Press.

[The title describes the contents of this important book].

Pais, A. (1994) *Einstein lived here*, Oxford: Oxford University Press.

[A collection of essays on aspects of Einstein's life. Complements *Subtle is the Lord* with new material that was not available in 1982].

Fölsing, A. (1997) *Albert Einstein: A biography*, New York: Viking.

[The most complete and up-to-date biography, but go to Pais for the science].

Index

aberration 87–92, 221–2
  light 87–90, 188–9
  stellar 91–2
Abraham, M. 50–54
acceleration 55, 57–62, 65–9, 76–80,
    128–31, 167, 176–80, 219
  Galilean transformation 69
  relativistic transformation 143–4, 179
accelerators 54–62
  colliders *see* collider
  Bevatron (Berkeley) 59
  Cosmotron (BNL) 59
  cyclotron 55–62
  electron 60–2
  Super Proton Synchrotron (CERN) 62,
    127
  synchrocyclotron 58–9, 62
  synchrotron 59–60
  Tevatron (Fermilab) 59, 61–2, 101
Adams, S. 239
Aitchison, I. J. R. 198
alpha particle 12
  decay 11–13
American Astronomical Society xii
American Institute of Physics 239
angular momentum 220
  spin 20, 163, 175–6, 197
annihilation 9
anti-matter 102
arrangement, standard *see* reference frame
astrophysics 82, 91–3
atomic mass 10–13
  unit, unified 16
  numbers 11

Bailey, J. 130
Barnett, R. M. (Particle Data Group) 61
basis vectors 214

Bell, J. S. 122
Berkeley *see* Lawrence Berkeley National
    Laboratory
Bernstein, J. 240
Bestelmeyer, A. 51
beta rays *see* electrons
binding energy
  atomic and nuclear 10–12
  deuteron 11
Bjorken, J. D. 196
BNL *see* Brookhaven National Laboratory
Boas, M. L. 19, 203, 208
Bohm, D. 147
Bohr atomic model 178, 193
boost, Lorentz 82–6, 93–96, 145, 147
    see also Lorentz transformation
Brace, H. B. 230
Brookhaven National Laboratory (BNL) 59
bubble chamber 31, 44, 94–5, 99
Bucherer, A. 51–4, 62

Cartesian *see* coordinate
Castor and Pollux 141
causality 116
centre-of-mass *see* kinematics
CERN xii, 37, 58, 60–3, 101, 106, 127,
    130, 132
chain rule 203, 209
charge, electric 47, 50–1, 70–1, 76, 100,
    230
  conservation 100, 199, 217
  continuity equation 199, 217
  current density 199
    four-vector 205, 207, 219
    transformation of 206, 216
  density 199, 205, 218

in **E** and **B** field 46, 48–9, 76–7, 198–200, 218
  invariance 76, 80, 201
Clock Paradox 128–31
clock 66
  ideal 66, 129, 131
  synchronisation 74–6, 146, 149
  zeroing 75
Cockcroft, J. D. 12, 15, 55
Colley, W. N. xii, xiii
collider 37, 60–3, 101–2
  centre-of-mass energy 60–1
  HERA (DESY) 61–3
  Intersecting Storage Rings (CERN) 101, 106
  Large Electron-Positron (CERN) 37, 60–3, 101
  Large Hadron (CERN) 61–3
  Tevatron (Fermilab) 61, 101
collision 6, 10, 13, 40–1, 58, 60, 63, 80, 94–5, 100–1
  see also scattering
composition of velocities 134–6
Compton, A. H. 35–8
  scattering 35–7, 185
conservation
  charge 100
  energy 5, 15, 112, 174–5
  energy and momentum 2, 4–5, 15, 17, 28, 36, 46
  four-momentum, validated 169–75
  law 67
  mass 5
  momentum 5, 24, 172
constant
  electric 70
  magnetic 70
  Planck's 7, 20, 87, 190
  speed of light 66, 81
continuity equation *see* electromagnetism
contraction *see* Lorentz–FitzGerald
coordinate 219
  Cartesian 18, 67–8, 83, 212
  spatial 69, 156
  spherical polar 214
  system 6–8, 152, 162, 214
  time 74–7, 146–7, 194
  transformation of 68–9, 109–11
cosmic radiation *see* radiation
cosmology 102
Coulomb (C) 10, 47, 53
Coulomb's law 70
covariant 67, 76

curvature 60, 99
  space 115, 212
cyclotron frequency 55–8

d'Alembertian operator 213, 217
Darwin, C. 141
de Broglie, L. 7, 182, 190–3, 198
decay
  alpha 11–12
  daughter nucleus 11–12, 20
  law of radioactive 126
  mean life 20, 26–8, 34, 65, 126, 128, 130, 146
  nuclear 12–13
  parent nucleus 11, 20
  particle 20–30, 99–100, 126–8
deuteron 11
Deutsches Electronen-Synchrotron (DESY) 61, 101
d'Inverno, R. 240
Dirac, P. A. M. 195
Dirac equation 182, 195–7
  covariance 195–6
  negative energy solutions 197
  positron 197
  spinor solutions 196
Doppler effect 78, 89–93, 189
  astrophysics 91–3
  longitudinal 89–90, 150
  relativistic 90–1, 189
  transverse 89–91
Drell, S. D. 196
dynamics 46, 59

$E = Mc^2$ 1, 4, 6, 11, 13
Einstein, A. xii, 1, 71–2, 74, 231, 239–40
  internet site about 239
Einstein
  electrodynamics 76–80
  papers of 1905 3, 46–7, 50, 72, 76, 78
  photon energy 7
  postulates 72, 75–6, 81, 111–12, 170, 198
  relativity 1, 72, 147
  summation convention 208–9
  railway train 73, 117
electric charge *see* charge
electrodynamics 46–7, 76–8

electromagnetism 1, 3, 46, 72, 182, 213, 215–9
  and Galilean relativity 70–2
  continuity equation 199, 217
  electric field 79
  electric current 70
  electrostatic potential 8, 206–7, 215, 218
  magnetic induction 19
  magnetic monopole, absence 218
  relativistic transformation of fields 76–7, 200–10
  scalar potential 215
  vector potential 206
  wave 1, 71, 188, 190, 199
electron 8–10, 18–19, 35, 48–54, 60, 62, 175–8, 221
  beta rays 51
  charge 8, 10, 51, 80
  discovery 18, 50
  e/m 50–4
  magnetic moment 176, 197
  rest mass energy 9
  spin 176, 197
  structure 50, 54
electron volt 8, 10
elsewhere 116
emission theories 235
energy
  centre-of-mass 39–41
  conservation 2, 5–6, 13, 15, 17, 28, 46, 174–5
  internal 14
  kinetic 5–8, 12–15, 20, 22, 48, 78–9, 83, 174
  negative 195, 197
  Newtonian kinetic 5, 14
  reaction threshold 42–3, 45
  relativistic kinetic 5, 78
  rest mass 6, 8–9, 14, 18, 39–40, 78, 174, 185, 192
  total 6, 14–15, 38, 166, 193
  units 8–10
energy-momentum *see also* four-vector
  transformation of 82–6
Epstein, L. C. 240
equivalence principle
  weak (WEP) 13
ether 1, 50, 54, 71, 76, 78, 90–1, 121, 219, 226–30
  detection of 227–30
  dragging of 232–3
  drift 227, 229
  lumiminiferous 226

European Organisation for Nuclear Research *see* CERN
event 66
  coordinates of 67, 76, 109–11, 145, 164,

Faraday tensor 202, 216
Fermilab 61
FitzGerald, G. F. 1, 230
fixed target 36, 40–1, 60–1
Fizeau 233
Fleming's left hand rule 47
Fölsing, A. 241
force 19, 51, 65, 167, 200, 218
  see also Minkowski force
  and acceleration 65, 69–70, 76–7, 79–80
  centripetal 47
  electric 46, 70
  Lorentz 47, 79
  magnetic 47, 70
  tidal 66
  unit, Newton 47
  work done 48
four-vector 17, 164–9, 219
  acceleration 167
  classification 169
  current density 205, 207
  electromagnetic potential 206
  energy and momentum 166–7, 182–7, 200, 221
  force 167, 200, 218
  kinematics with 182–7
  null 169
  time–space 109–11
  velocity 136, 165–7, 174
  wave 167–8
  transformation of 164
frame *see* reference frame
French, A. P. 239
frequency, proper 87
Fresnel 222
future 115–16, 136

Galilean relativity, principle of 70, 72
Galilean transformation 68–71, 83–5, 110, 146–7, 157
  invariance under 70
Galileo Galilei 64, 226
gauge
  invariance 207
  Lorentz 207, 217
  transformation 206
geometry, differential 215
Goldstein, H. 178
Gordon, W. *see* Klein–Gordon equation

gravity 13–14, 66, 130, 179
  centre of 40
  field 130
group velocity 7, 166, 188–9, 191–2
gyroscope 175

hadron 96
Hagedorn, R. 102
Hall, D. B. 127
Harris, S. xii, xiv
Hawking, S. 1
Hecht, E. 229–30
Heisenberg, W. C. 27
  uncertainty relation 27
Hey, A. J. G. 198
Hubble Space Telescope xiii
hydrogen atom 197

inertial frames *see* reference frames,
  inertial
  equivalence of 70–2, 145, 147
  standard arrangement 68, 75, 82–3,
    109–10, 145, 155, 157–9
Institute of Physics xii
interferometer, Michelson 227–9
interval 66, 115, 131, 209–10, 212
  space-like 115, 140, 169
  squared 115, 169
  time-like 115, 140, 169
invariance 71–2, 84–5, 146, 183
invariant 66, 221
  form 66
Ives, H. E. 78

Jackson, J. D. 178, 207, 235, 239
JINR (Joint Institute for Nuclear Research,
  Dubna) 58

Kaufmann, W. 51
KEK (National Laboratory for High
  Energy Physics, Japan) 102
Kennedy, R. J. 231–2, 235
kilogram 8–10, 53
  defined 81
kinematic quantities, table of 14
kinematics 4, 17, 27, 35, 46, 99–100, 163
  hybrid 13
  relativistic 12, 17–45, 82–102, 182–87
    centre-of-mass 37–43, 60–2, 86,
      100–1, 184
    decay, moving particle 28–31
    formation 31–5, 42, 102, 106
    three-body decay 24–8
    two-body collisions 41–3, 101–2

    two-body decay 20–4
    scattering 35
    with four-vectors 182–7
    zero mass particles 35
Klein–Gordon equation 182, 194–5, 197–8
Klein, O. *see* Klein–Gordon equation

laboratory frame 32
Lagrangian 220
Larmor, J. 1
Lawrence Berkeley National Laboratory
  (Berkeley) xii, 56, 58–9
Lawrence, E. O. 56–8
Lewis, G. N. 80, 167, 170
LEP *see* collider, Large Electron-Positron
LHC *see* collider, Large Hadron
life, mean 34, 65, 126, 128, 130, 141–2,
  146
lifetime, average *see* life, mean
light
  cone 115–6
  propagation 73, 88–9
  speed of *see* speed of light
limiting speed *see* speed of light
line-of-sight 88–9
Livingston, M. S. 56
Lorentz, H. A. 1, 50, 71–2, 78, 231, 240
  electrodynamic theory 231, 234
  factor 7, 14, 39, 41, 100–1, 152
  force 47, 79, 200–2, 219
  mass variation with speed 49–51, 76
Lorentz–FitzGerald contraction 119–21,
  132, 139–40, 144, 230–1
Lorentz transformation 43, 75–6, 84–6,
  109–12, 145–6, 157
  see also boost, Lorentz
  derivation 152–6
    using conventional method 152–5
    using k-caculus 147–152
    transverse coordinates 155–6
  general 158–9, 179
  graphical representation of 96–9
  in particle decays and collisions 99–101
  non-commuting 160–3, 176
  of energy-momentum 82–6, 93–4, 109
  of time–space coordinates 109–11
  successive boosts 159–63
Lyons, L. 203

McGuinness, M. 240
McMillan, E. M. 58
Martin, B. R. 157, 240

mass
    conservation 5
    gravitational 13
    inertial 13
    invariant 40
    longitudinal 77
    missing 186–7
    point 67
    relativistic 49–51
    rest 18
    rest energy 6, 8–9, 14–15, 18, 39–40, 78, 174
    transverse 77
    units 8–10
    variation with speed 49–51, 76
matter, excess over anti-matter 102
matrix, symmetric 159
Maxwell, J. C. 1, 51, 71, 199, 226
Maxwell's equations 3, 19, 198–9
    covariance of 71, 76, 204–6, 218
meson 20
Michelson, A. A. 50, 227, 229
Michelson–Morley-experiment 121, 227–9, 235
Minkowski, H. 72, 136–7, 139, 144, 231, 236, 240
    force 167, 218
    map 136–40, 144, 148
Misner, C. W. 215
Møller, C. 131
momentum 5, 46–50, 77, 79
    centre-of- 38
    conservation 5, 24, 80, 167, 175, 183
    definition of Newtonian 5
    definition of relativistic 5, 79
    four 182–3
    longitudinal 77, 93–4, 142
    transformation of 96–9
    transverse 77, 93, 95
    units 9–10
    wavelength 7, 190–2
Moore, W. 193
Morley, E. W. 50, 121, 227, 229
motion, circular 47–8, 50, 78
moving object, appearance of 122–3

Nature (Journal) 130, 141
Newton, Isaac 49, 65, 69–70
    and Galilean transformation 70
    assumptions 68–69
    laws of motion 65, 70, 72, 76, 79, 112
    momentum 20
    unit of force 47, 81

universal time 68, 72–74
Noble, H. R. 230
nucleus, atomic
    binding energy 10–11
    decay 12, 13
    reaction 12, 15

observer 64
    inertial 67, 73–4, 88, 130, 146
operator, differential 193–5, 199
    transformation of 203, 213
    vector 19, 150, 199, 205–7, 216–17
operator-observable 194
Organisation Européenne pour la
        Recherche Nucléaire *see* CERN

Pais, A. 198, 241
Panofsky, W. K. H. 229–30
particle 4, 67
    anti-particle 197
    decays 20–31
    energy and momentum 5–6, 18, 79
    physics 18, 20, 26, 59, 102, 174
    secondary 94–5
    speed 15
    velocity 18
particles named
    B-meson 102, 141–2
    charged pion $\pi^{\pm}$ 10, 20, 22–4
    D-meson 26–8
    electron *see* electron
    first artificial pion production 58, 62
    Higgs boson 62
    kaon 26–8, 30
    lambda Λ 30
    muon μ 20, 22–4, 127–8, 130–2
    neutral pion π^0 10
    neutrino ν 22, 24, 35
    neutron 11
    positron 9, 24, 60, 101, 106, 197
    quark 96
    photon *see* photon
    proton *see* proton
    upsilon Υ 102, 106, 141–2
past 115–16, 136
Pauli, W. 1, 197
    exclusion principle 197
Perkins, D. H. 100
permeability of free space (μ_0) 70, 81
permittivity of free space (ε_0) 70, 81
phase stability 58–9
photon 7, 10, 14, 35–6, 87–92, 98, 113, 192, 219
Phillips, M. 229–30

Planck, Max 7, 77, 79, 202
  constant 7, 10, 20, 190
  definition of momentum 77, 79, 202
  photon energy 7
Poincaré, H. 1, 171
Poisson's equation 218
positronium 9
Powell, C. E. 23
product
  direct 211
  inner 19, 164–5, 167, 212, 215
proper quantities 65, 146
  length 119, 120–1, 131, 143
  frequency 87
  time 131–2, 165, 179, 200, 218
proton 8, 10–12, 15, 19, 55, 57–9, 61–2, 221

quantum
  chromodynamics 96, 236
  field theory 236
  mechanics 2, 182, 192–8
Quinn, H. R. 102
Q value 11–13

radiation
  cosmic 19, 23, 58, 127, 197
  synchrotron 60
radioactive decay law 126
rapidity 94, 105
Rayleigh, Lord 230
reaction threshold energy 42–3, 45
reciprocity 146
reference frame 18, 37–40, 65, 67–8
  inertial 65–72, 74–8, 82, 87–8, 93–4,
    109–10, 129, 145–9
  instananeously comoving 176–7
  rest 18, 146
  standard arrangement 67–8, 82–3,
    109–10, 140, 145, 155–61
refractive index 222
relative motion 18, 25, 40, 64–6, 85
relative speed of inertial frames 78
relativity
  formalities 145–78
  foundations of special 64–81
  general theory of 4, 66, 130, 208
  principle of special 64
  special 1–3, 5–6, 49, 55, 59, 182, 198
representation, graphical *see* Lorentz
  transformation
resonances 33
rigid body 67
Rindler, W. 239

Rosser, W. G. V. 147, 239
Rossi, B. 127
rotation 66, 163, 175–6
Rutherford, E. 51

Sard, R. D. 240
Sartori, L. 240
scalar 164–5, 194, 213
  function 206
  product 19
  wave function 194
scattering 35, 38, 170
  Compton 35–7, 185
  cross-section 33–4
Schrödinger, E. 192
  wave equation 192–5
Schutz, B. F. 215, 240
Schwartz, J. 240
Schwartz, M. 199
Scott, G. D. 122
Shaw, G. 157, 240
SI *see* Systèm International d'Unités
SI unit
  length 66, 81
  mass 81
  time 66, 81
simultaneity 72–4, 117–19, 140
  absence 117
SLAC 15, 102
space
  homogeneity 146
  isotropy 146
space–time *see* time–space
spectrum, electromagnetic 89, 223
speed of light 1, 4–6, 17, 60, 66, 71–8,
  111–12, 145–7, 227
  exact number 81, 146
  fractional 7, 15, 21, 32, 41
  limiting 59–60, 62, 78, 116
speed, relativistic 18
spin 20, 163, 175
Stachel, J. 78, 240
Stanford Linear Accelerator Center
  see SLAC
Stewart, A. B. 91
synchrocyclotron 58–9
synchronisation, clocks 75–6
Systèm International d'Unités (SI units) 9,
  10, 66, 71, 88

target, fixed 36, 40–1, 60–1, 63
Taylor, E. F. 64, 147, 239

tensor
  contraction 209
  contravariant 208
  covariant 208
  differential operators 213
  differentiation 213–4
  dimension 208
  direct product 211
  dummy index 209
  Faraday 202, 216
  formulation of electromagnetism 215–9
  index 202, 208, 213, 225
    lowering 212
    raising 212
  metric 211, 212
  mixed 208, 210
  null 211
  rank 208
  rank-0 211, 213
  transformation 210
  type 210
Tensorland 208–15
Tevatron (Fermilab) 59
Thomas, L. H. 163
  precession 175–8, 197
Thomson, J. J. 50–1
  discovery of electron 18, 50
Thorndike, E. M. 231–2, 235
Thorne, K. S. 215
thought experiment 73
time dilation 91, 102, 122–5, 127–32,
    139–40, 162
  experimental tests 125–8
time ordering 115–6, 140
time–space 3, 109
  transformation of 109–111
Tolman, R. C. 80, 167, 170
Torretti, R. 78
transformation 67
  Galilean *see* Galilean transformation
  gauge 206
  improper 157
  inhomogeneous 157
  Lorentz boost *see* boost
  Lorentz *see* Lorentz transformation
  matrix 86, 113
  proper 157
  reciprocity 84
  rotation 157
  tensor 210
Trouton, F. 230
Turner, E. L. xii, xiii
Twin Paradox 128–31

twin, travelling 128–9
  and muon storage ring 130–1
Tyson, J. A. xii, xiii

unification of electricity and magnetism 71
uniform motion 64–5, 146
units 8–10, 47
University of Birmingham 59
  of Chicago 58
  of Liverpool 58

Van Driel, H. J. 122
vector field 214, 217–18
vector meson 102
vector notation 14, 18–9, 21, 29, 182
  product 19, 47
Veksler, V. I. 58
velocity
  fraction of *c* 22, 200
  group 7, 166, 188–9, 192
  mutual 134
  phase 7, 167, 168, 188, 192, 222
  relative 134
  transformation of 69, 83, 132–6, 141,
    166
vertex 29, 99

wave
  equation 192–5, 207
  four-vector 167, 168, 188, 190–1
  frequency 191
  front 168, 188, 190
  function 193–6
    scalar 194
  group velocity 166
  number 168
  phase 168
  phase velocity 191
  shock 222
wavelength 7, 35, 37–8, 65, 190
Weinberg, S. 209
Weyl, H. 72, 231, 240
Wheeler, J. A. 64, 215, 239, 240
Witherell, M. S. 102
world line 113–16, 131–2, 136–9, 147–51,
    165
  photon 113–4
Walton, E. T. S. 12, 15, 55
Woodgate, G. K. 178

X-ray 35–38

Zajac, A. 229

DATE DUE